# Dams in Africa

# Dams in Africa

## AN INTER-DISCIPLINARY STUDY OF MAN-MADE LAKES IN AFRICA

EDITED BY
NEVILLE RUBIN
AND WILLIAM M. WARREN

FRANK CASS & CO. LTD.
1968

First published in 1968 by
Frank Cass and Company Limited
67 Great Russell Street, London WC1



SBN 7146 1248 O



Printed by W. & G. Baird Ltd., Belfast

# CONTENTS

## LIST OF MAPS

# PREFACE

THE present volume originated in a seminar which was conducted at the School of Oriental and African Studies of the University of London during the 1965–66 academic session. It represents part of a continuing process which seeks to involve the different Departments of the School concerned with African Studies in the advancements of inter-disciplinary work in this area.

The choice of dams as the subject for such an inter-disciplinary exercise was a natural one. There can be few other topics which provide a more convenient and meaningful rendezvous for the various academic disciplines whose activities relate to the manifold problems of African development, and which make it possible for material of such considerable practical importance to be viewed on a comparative basis by specialists concerned with different parts of Africa. The significance of dams for any consideration of the problems of African modernisation hardly needs to be emphasised. With the possible exception of railways, there have been no projects in Africa of comparable size and implications to the giant dams, involving as they do such unparalleled commitment of resources, with corresponding social, political and legal consequences. Some of these consequences have been considered in the essays presented here, both in the specific context of particular projects and as in the more general context of comparative surveys, by experts who were (with two exceptions) present at the seminar. In the first of these, Professor Harrison Church provides a general introduction to the subject, which surveys not only the geographical factors involved in the construction of dams, but also indicates the nature of the economic and developmental considerations which affect choice of sites and types of dams. There follows a discussion by Mr. Joy of the role of the economist in the complex of decisions connected with dam-making which draws on the author's experience of such projects in both Asia and Africa. Drs. Brokensha and Scudder, using material gathered from studies made in relation to dams constructed in Nigeria, Ghana and Zambia, adopt a comparative approach in their essay on the problems of population re-settlement, and the degree to which these were foreseen and solved. The next two essays deal specifically with the Volta River Dam project in Ghana; Professor Steele provides a case-study of the project from the viewpoint of a geographer, while also bringing it into historical and economic focus; and Mr. Rouse Jones provides a unique survey of the legal machinery which was set up to ensure adequate safeguards for those who made available the considerable funds which the project entailed. Next, Professor Elkan and Miss Wilson offer a further case-study—this time, from an economist's standpoint—of a more limited project, the Owen Falls Hydro-Electric Dam in Uganda. There follow three essays which deal with the political implications of dam-building in Africa. Dr. Watt considers the historical antecedents, as well as the political ramifications, of the decision to construct the High Dam at Aswan. Dr. Simons deals with the geo-political, as well as the economic, advantages and disadvantages of the plans to dam the waters of the Orange River in South Africa. And Mr. Austin describes the problems of international law which have resulted from the erection of the Kariba dam in the former Central African Federation now that the Federation has been broken up. The volume concludes with a critical assessment, by an economist (Mr. Stanton Smith of the Kaiser Corporation), of the actual and potential hydro-electric resources of West Africa, both Anglophone and Francophone.

It is our hope that the appearance of these essays will draw attention to the importance of the experience gained from dams and other projects of a like nature in the study of the whole process of modernisation in Africa. It is, indeed, surprising that no systematic attempt has been made to assess the experience of African dams, particularly in view of the diversity of discussion around a common theme which has characterised their planning, construction and utilisation. The subject is not only of obvious interest to policy makers in view of the scale of human, natural and financial resources involved, but has also aroused considerable public curiosity and enthusiasm (as well as concern) as a result of the sheer physical size of some of the projects and the signal engineering achievements which many of them have entailed. This interest has already stimulated a growing popular literature on dams[1] which has not, however, been matched by a corresponding expansion of academic work in this field.

The gap in the literature here is, of course, representative of a more general weakness in writing displaying an inter-disciplinary approach to work on African themes. There has been, in recent years, a proliferation of publications on numerous aspects of African development and change, much of which is specifically multi- or inter-disciplinary in character. On the whole, however, this material has been either of a general nature and wide-ranging in character[2], or owes its less general focus to the more or less random collision of economists, political scientists, sociologists and anthropologists working within the confines of a particular country or region of the continent[3]. These collections have ther own merits but clearly the full value of a multi-disciplinary approach will not be realised without a greater concentration of effort on more narrowly defined topics. There are, of course, exceptions to the general weakness here, among which the work on problems of urbanisation and urban life[4] is of course outstanding. Yet these are exceptions and there are several topics which have, for some time, been calling out for inter-disciplinary treatment: to take but three examples, labour relations, railways, and agricultural extension would benefit greatly from a collaborative study by those approaching them from complementary fields of expert knowledge.

The papers contained in this volume do not, of course, purport to contain the results of any such effort at collaboration. Neither do they represent a piece of inter-disciplinary research. Instead, they provide a confrontation of disciplines around a single topic, which is intended to stimulate further work in this field and to provide some insight into the advantages of multi-disciplinary cooperation. Thus, while the dominant theme of many of the papers is critical of one or other aspect of dam building, it is of far greater importance that, together, they illustrate the benefits of adopting an all-round approach. Two conclusions emerge. The first is that it is apparent that, in future, all major decisions concerning African dams should be taken in the light of information provided through employing the skills of considerably more disciplines than have hitherto been involved in any single project. And the second is that the practitioners of the various disciplines should themselves become increasingly aware of the relevance and findings of others working in different fields, but who are equally involved in a total human situation. If such an understanding is in any way advanced by the present volume, the editors and contributors will be well satisfied.

The editors wish to acknowledge with grateful thanks the help of many colleages and participants in the seminar; to thank those who contributed papers to the seminar, as well as Drs. Brokensha and Scudder, and Mr. Smith, who were kind enough to submit papers not originally available to the participants; and to express their gratitude to Professor A. N. Allott, Professor of African Law in the University of London, for initiating the series of inter-disciplinary seminars at the School of Oriental and African Studies.

N. R.

W. M. W.

[1] See Leslie Greener, *High Dam Over Nubia,* 1962; David Howarth, *The Shadow of the Dam,* 1961; Tom Little, *High Dam at Aswan,* 1965.

[2] W. R. Bascom and M. J. Herskovits (lds.), *Continuity and Change in African Cultures,* 1959; M. J. Herskovits and M. Harwitz (eds.), *Economic Transition in Africa,* 1964; and R. A. Lystad, *The African World: A Historical and Political Survey,* 1965, are good examples of this.

[3] See, for example, the excellent collections *Political and Social Change in Modern Egypt,* 1968, and *A Study of Contemporary Ghana,* 2 volumes, 1966, edited by P. M. Holt and W. Birmingham, I. Neustadt and E. N. Omaboe respectively. Of a somewhat different character is Ronald Segal and Ruth First (eds.), *South West Africa: Travesty of Trust,* 1967.

[4] Thus, supplementing a number of useful individual studies, we have Hilda Kuper (ed.), *Urbanisation and Migration in West Africa,* 1965; P. C. Lloyd, A. L. Mabojunge and B. Awe (eds.), *The City of Ibadan,* 1968; and Horace Miner (ed.), *The City in Modern Africa,* 1963.

CHAPTER I

# A Geographical View

## R. J. HARRISON CHURCH

FEW of man's modifications of the landscape can initiate such profound physical, economic and social changes as dams. They regulate and enlarge rivers to make them serve man's general thirst for water, or his more specific needs of irrigated agriculture and power. In most cases there are also subsidiary benefits in, for example, health, navigation and tourism.

There is some confusion corncerning the terms 'barrage' and 'dam'. The former is rarely used in the United States of America, but elsewhere it denotes a structure which does not provide complete water control, as does a dam, but raises the level of a river to such a height that the irrigable area is 'commanded' by (i.e. can all be watered from) the new level of the river. A dam may store water to its capacity, and may also produce power, permit irrigation, or serve other needs.

Africa presents certain advantages for the construction of dams; it also has some severe disadvantages and problems, although most of these apply equally to all tropical or equatorial areas. Africa is, in general, an ancient continent, and so is composed of some of the world's most ancient rocks. These have been subjected to many cycles of uplift and erosion. Africa is, in consequence, a continent of plateaux, each with usually well-marked edges. Rivers fall over these, and such sites may provide points for power dams, as they have in western Angola, as they could do in the Kouilou gorge of the Niari river in Congo (Brazzaville) or in the long gorge of the lower Congo, especially in the Inga rapids. It is said that the power potential of these is equal to all that so far developed in the U.S.A., or to over 40 per cent of all world power potential.

Another relevant geological phase was that of the mid-Cretaceous to Tertiary period when the rift valley system was formed. Within its valleys are many rivers and lakes, while the rifting also caused the diversion eastwards of the Zambesi[1]. The continent is so old that the drainage pattern has inevitably been subjected to many changes by capture or diversion. Many rivers are plainly 'multiple' ones e.g. the Nile, Niger, Congo and Zambesi. As such, they inevitably have many breaks of slope, or sites for possible dams or barrages, such as the Portes d'Enfer of the upper Congo. These rapids lie at the point of capture of the upper Congo (or Lualaba) by the middle Congo, when the Lualaba could no longer flow into the Nile system, consequent upon the development of the rift valley system and its associated volcanism, which blocked the outlet. Africa is not short of suitable sites for dams, although many are in rather inaccessible places. Rather does Africa lack the technological, economic and political conditions to encourage the construction of dams.

Again because Africa is so old, it has many hard and largely impermeable rocks. These may 'rot' superficially and so give rise to constructional problems (as with the Guma Dam, south of Freetown, Sierra Leone), but such rocks usually provide a firm 'anchor' for a dam and good building material. On the other hand, run-off is rapid, and so rivers rise quickly and with dangerous force, the more so as rainfall is mostly seasonal in Africa.

Likewise because of the hardness of most African rocks, the rivers (the Nile excepted) carry little silt, so that siltation is not a problem, except for the Nile dams. It is especially the Blue Nile which carries so much fine silt from the younger and softer volcanic rocks of Ethiopia, which there cover the ancient basement complex. Silt accumulation in Nile dams reduces their life and storage or power capacity, while the silt is lost to farmers. To counter it, the old Aswan and the new High Dam store water only after the first floods (which are heavily silt-laden and refertilise certain areas) have passed. With the completion of the Aswan High Dam, the Nile dams will be holding back some 10 of the 13 million tons of silt Egypt previously received annually[2], but that is part of the price for vastly enlarging the area, period and effectiveness of irrigation. Fertilisers (which will be made more cheaply by power from the Aswan High Dam) can make good most of this loss.

The Congo, alone of the major rivers, has a remarkably even régime, with a flow ratio of only 1: 3. This is because the river and its major tributaries straddle the equatorial zone, with a fairly evenly distributed rainfall. The White Nile has a similar character; it is the Blue Nile that causes the wide fluctuations of the Nile proper.

Most African rivers have very uneven régimes, so that their

depth and speed of flow are subject to great fluctuation, like normal tropical rivers. Thus the flow ratio of the Zambesi is 1: 13, and that for the Nile in Egypt 1: 17. So dams must necessarily be high to form large retaining lakes to regularise flow whatever the season. On the other hand, such large expanses of water are subject to heavy losses by evaporation, especially in South-Central Africa, the Sudan, Egypt and North-West Africa, where water is most precious.

The rift valley areas of Africa, and also the southern Cameroon, Accra (Ghana), Dakar (Senegal), Moroccan and Algerian areas, as well as most African islands, are prone or could be prone to earthquakes, so that dams in these areas are subject to this hazard. To minimise it, dams in such areas should have a clay core with rock fill of increasing size on either side; the Volta Dam at Akosombo is of this type[3]. On the other hand the Kariba Dam is of the concrete arch type, although it lies within an area of faults[4]. 'One such major fault runs parallel with and near to the bed of the river in the vicinity of the dam site. The country flanking the river on both sides is traversed by a complicated pattern of faults. There is conjecture as to what effects may be caused by the loading of the earth's crust at this site, with the weight of the enormous volume of water contained in the Kariba lake. It is known, for instance, that when the Boulder Dam was built in the U.S.A. the weight of water caused the earth's crust to "bend". At Kariba, the Director of Federal Trigonometrical and Topographical Surveys has initiated a study of the behaviour of the earth's crust'[5].

In its early days Lake Kariba experienced severe growth of *Salvinia auriculata,* and of *Pistia stratiotes* which upset fish and impeded fishing and boats, as it has on the Congo River. At Kariba the problem has greatly diminished as former vegetation in the lake area has decayed and been removed, so increasing the movement of wind and water on the lake surface. Rather more care was taken in the prior removal of vegetation in the valley of the Volta in Ghana.

### THE PURPOSE OF DAMS

There are the fairly obvious technical purposes of dams—headwater and flood control (important on the Nile and Orange rivers), water supply for man and animals (of increasing importance in dry Africa), for irrigation, power, or a combination of one or more of these. Governments are usually interested in other major effects of the construction of dams, such as the acquisition of new labour skills, and the diversification of economies and standards of living after the dams are completed. These can occur in industry, in new methods and products of agriculture, in subsidiary benefits such as transport, fishing and recreation and, not least, in the improvement of existing production. Governments must decide whether money spent on building a dam is better used than on some totally different development of its environment e.g. transport, whether one large or several small dams are to be preferred, and whether expenditure is to be strictly economic or partially or wholly regarded as expenditure on social welfare. Some African governments also view dams as prestige projects, to be numbered with an embassy in Washington.

The cost of a dam will be determined very largely by the site and its geological character. Accessibility will greatly affect it; thus the Kariba and the Kainji (Nigeria) dams were built in remote areas, while the Volta and Owen Falls dams were constructed at accessible sites. The character of the river bed and valley sides will largely determine the amount of excavation required, and the type of dam which may be built, while the régime of the river will greatly affect the nature and amount of diversion and control works required to permit construction of the dam.

A decision must also be taken as to whether a dam shall be multi-purpose or not. This is obviously the ideal, but the costs may not justify it. Water impounded for power from one dam is usually needed throughout the year, especially in the early schemes, while irrigation water may be required only seasonally. Many of the best sites for power production (as in Angola) lie at the limit of very narrow coast plains, where the possibilities for irrigation are few. They are also minimal in the gorge immediately below the Kariba Dam, rather more promising but not ideal in the Accra Plains below the Akosombo Dam on the Volta, and considerable only in the wider coastal plains of Mozambique, South Africa and Morocco. It may sometimes be possible to add a purpose or facility to a scheme; thus the Sennar Barrage was originally built for irrigation but has since been raised and otherwise modified to produce power as well.

Without being multi-purpose in design a hydro-electric scheme may have varied consequences, since its power can be used to run irrigation pumps, improve the productivity of agriculture, develop agricultural processing, mining, industry, transport and other activities. To a lesser extent this is also true of irrigation or water reservoir dams, since they may greatly improve the efficiency and variety of agriculture, but the effects are obviously narrower in scope if not also in monetary terms.

As well as the possibility of constructing multi-purpose dams, we may note the increasing tendency towards the comprehensive development of valleys under one authority, such as the Volta River Authority, the Awash Valley Authority, and the Orange River Development Scheme. Furthermore, the Niger has been the subject of an international convention since 1885, which was replaced by another in 1963. The Gambia and Senegal valleys, and the Chad Basin, are now the subject of such agreements[6], while the Nile is governed by the Egypt-Sudan Nile Waters Agreement of 1959, which replaced the Anglo-Egyptian Nile Waters Agreement in 1929. The Zambesi is likewise governed by international agreements.

#### DAMS FOR IRRIGATION

Most African barrages and dams have been built for the needs of irrigation, although the famous dams of recent years have been built mainly for power. The Nile still has the greatest number of dams and barrages of any African river, because of pressure on land in Egypt and partly because many of the Nile's riverine peoples have long been accustomed to the use of its waters for irrigation—albeit by simple means.

The earliest barrage schemes (the first was in 1861) were designed to extend the season of irrigation from the months of flood-retreat to those of mid- and late-summer. These barrages had to be supplemented by dams if at first early summer, and, still more, if all-year irrigation was to be provided. So the first Aswan Dam was constructed in 1902, raised in 1912 and 1934, and is now being largely supplanted by the Aswan High Dam four miles upstream. Where only one crop could formerly be grown in Egypt, three or four may now be raised annually, and this over an area many times larger than before. Egypt and the Sudan have been utterly transformed; between 1913 and 1963 there was a 100 per cent increase in Egypt's population and a 20 per cent increase in the irrigable area. The Aswan High Dam should achieve another such transformation. Another two million acres will become available for irrigation, while power resources will be trebled to the great advantage of new industrial development and location, and cheaper and additional mining.

Meanwhile, the Sennar Barrage had been completed in 1925 on the Blue Nile to irrigate a million acres in the Sudan's Gezira, although the barrage also regulates that river after its annual peak. The Jebel Aulia Dam was finished in 1937 in the Sudan, but on behalf of Egypt, to store water by holding back most of the White Nile to make way for the peak floods of the Blue Nile. In 1954 the Owen Falls Dam, although built for power production, also converted Lake Victoria into a vast reservoir for Egypt —and that at her expense. Similar dams could be built across or near the outlets of lakes Albert and Tana. The Sennar Barrage, now raised and also producing power, is supplemented by the Roseires Dam which is making possible another 800,000 acres of irrigated land in the Managil, and 1,200,000 acres in the Kenana area to the south of the Managil and Gezira. Even more recently the Khasm el Girba Dam makes possible the irrigation of some 500,000 acres in that area near the Atbara River, and the resettlement of Nubians from the area flooded to the south of the Aswan High Dam.

These details have been cited because of the vital character of irrigation to the Sudan and Egypt (which also use other methods of irrigation—traditional and modern), and because of the contrasted régimes of the fairly even-flowing White Nile and the sharp annual floods of the Blue Nile, Atbara and Sobat. Great advances have also been made in Morocco, Algeria, Tunisia and the Republic of South Africa, although in all these white settlers or farmers were the original and main beneficiaries. Only in the latter country was there another major river capable of comprehensive development—the Orange—but of small scale compared with the Nile, and the subject of Chapter VIII of this book.

Governments should decide what methods of farming shall be undertaken on an irrigation scheme, whether they shall be mechanical with paid labour, mechanical with peasants, non-mechanical with peasants, or a combination of these in different areas of the same scheme. If peasants are to be introduced, they may already be skilled farmers, or they may need subsidisation and training while they cultivate. The size of their holdings must be decided—a very difficult yet crucial matter. Crops, crop combinations, rotations, methods of harvesting, processing and marketing call for decisions between alternatives, and conflicting and sometimes uncertain advice in the absence of sufficient research or records.

The Gezira is a rare example of where the land-owning system has been reformed by retaining under a tenant system some of the original owners. To these have been added far more tenants from farther afield, most with some idea of the use of irrigation water.

Control has been close, though in a partnership, and markets have been generally good for cotton, the main cash crop.

Certain schemes have had a prominent social welfare aim, as at Mwea Tabere in Kenya initiated for Mau Mau detainees, and now operated for landless farmers from over-crowded areas. In the Medjerda Valley of north-eastern Tunisia the large estates of colonial days, farmed extensively, have been broken up to receive poor settlers from over-crowded southern Tunisia, who have to be taught to farm intensively with irrigation. The Inland Niger Delta scheme in Mali originally attracted farmers from the over-populated Upper Volta, but they now find themselves in a country politically unfriendly to them. In Angola and Mozambique resettled Africans are outnumbered by, but may live and work side by side with, poor immigrant Portuguese.

In all schemes there will be new and radically different settlement and housing patterns, and new medical problems such as the likelihood of a greater incidence of Bilharziasis. Soils may become toxic through lack of or inadequate drainage. On the other hand, barrages and dams for irrigation have permitted the introduction of new or far more remunerative cash crops such as cotton to Egypt and the Sudan, and to the Danakil Desert of Ethiopia; of citrus and sugar to the Lundi Valley of Rhodesia; of sugar at Kilombero and Arusha Chini in Tanzania, and at Wonji in the dry Rift Valley of Ethiopia; of artichokes, tomatoes and grapes for export from the Medjerda Valley of Tunisia. In turn, these have permitted new processing and manufacturing industries of textiles, sugar refining, canning, jams and confectionery.

Irrigation from dams requires great organizational and managerial skills that are rare in Africa, but if such improved agriculture can be achieved the results may make a greater impact on the rural African than power dams, which tend to help mainly the industrial and urban population.

### DAMS FOR POWER

The decision has rarely had to be taken in Africa as to whether power should be generated thermally or hydraulically, for coal is usually either rare or remotely situated, whilst oil and natural gas are so far being used in few places for the generation of power. Only in the Republic of South Africa and in Nigeria have there been realistic alternatives between thermal and hydro-electric power. In the first case, this was largely because of the abundant and cheaply mined coal on the one hand, and highly seasonal and low rainfall on the other hand, of much of that country. In Nigeria there is abundant oil and gas in the Niger Delta and its confines, but the reserves of these were insufficiently known when a decision had to be taken on the Kainji Dam[7]. The ultimate intention is that there shall be a national electrical grid, with its northern part largely fed by hydro-electric power from the Niger Dams Project, while the southern part will receive power mainly or entirely generated from oil and natural gas.

The cost and difficulties of rail haulage of coal were powerful incentives to the building of power dams on the upper waters of the Congo, to serve the many and varied mines of Katanga. These incentives were even stronger to the construction of the Kariba Dam to dispense with the dependence by the Zambian Copper Belt upon coal from the Rhodesian collieries at Wankie, and the long haul from the latter to the former. Power from the Owen Falls Dam is also used in copper smelting, but in this case at the site of the power, to which a 26 per cent concentrate is brought 266 miles by rail from the mines at Kasese in western Uganda[8].

Both the Volta Dam and the Sanaga Dam at Edea in Cameroon provide power mainly for the smelting of imported alumina to make aluminium ingots which, in turn, are for export. However, the former was built with this aim mainly in view, whilst this function was somewhat of an afterthought at Edea. It, like Owen Falls, was rather an act of faith, a case of building a dam and then attracting industrial and other users. This has not proved easy at Owen Falls, where potential power is still not fully developed.

Dams should be built only when it is clear how the power can and will be used. It is also axiomatic that the users should be as numerous and varied as possible, to offset short and long-run changes in demand. Power from both Kariba and Owen Falls goes to two countries in each case, and to a fair diversity of users. However, in the case of Kariba, the political problems have been acute; while the use of power is varied only as a whole and not in each country, since in Zambia it is used largely in copper-refining, but in Rhodesia it is used mainly in industry. In Uganda the uses are more varied but few, and insufficiently developed; whilst in Kenya power is taken mainly by industry, there are complaints about the charges, and so a strong nationalist urge exists to develop local power potential on the Tana River, despite the high cost this entails.

Power installations are subject to greater risk in the tropics compared with temperate lands, because of the higher relative humidity in the rainy season and the frequency of severe electrical storms. To counter these raises the cost of both installation and maintenance. Domestic demands, already small in such poor countries, are not easily stimulated because electrical wiring is difficult in mud houses. Nevertheless, cheap power, like cheap and efficient transport, can be a major key to general economic development.

### THE RESETTLEMENT OF DISPLACED PERSONS

Dams raise the natural level of rivers or lakes, so flooding lands previously settled, farmed or periodically grazed. Fortunately, the flooding is often in remote and thinly-settled areas, especially the usually dry regions upstream of irrigation dams. Yet the huge lakes formed by the Kariba, Volta and Aswan High dams forced the resettlement of between 50–80,000 people in each case. At Kariba 51,000 of the rather retarded Tonga and We people have been moved to higher levels[9]. There they have become interested in fishing, which they previously neglected because the river flowed too fast, they have also improved their dry farming, and have higher standards of living and of health.

In the valley of the Volta and its tributaries some 78,000 people had to be moved from 15,000 homes and 740 villages. They are now living in 52 new villages, all carefully sited and planned, and varying in size from a few hundred to several thousand inhabitants. All have, or will have, sewage disposal and piped water. Great trouble was taken to ensure that kin and tribal groups were kept together. All are being encouraged and guided to do group farming and marketing. For most there is the prospect of irrigated farming, fishing and trade on the new water highway. All have better homes or the chance and compensation payments to enlarge their government-built free one-room houses, which have concrete bases permitting the construction of additional accommodation. Resettlement had to be carried out more hurriedly than expected, and some anticipated water supplies proved inadequate, but the long-term prospects are incomparably better than they were for these people in their original villages[10].

The movement of the Nubians disturbed by the Aswan High Dam has been more complicated. They lived mainly within the Sudan, in the Wadi Halfa area, and have traditionally felt neglected by Egypt and the Sudan. With Egyptian compensation they have been moved from their intensely arid but fairly healthy environment to the much less dry and less healthy Khasm el Girba irrigation project some 600 miles to the south-east, where the environment is utterly different, as are the lives of these displaced people.

## DAMS AND OTHER FACILITIES

The need for access to dams under construction and to import the necessary heavy equipment has notably improved general communications. Thus the Khartoum-Sennar railway was built in 1911 in anticipation of the construction of the Sennar Dam, and of the traffic which its irrigated lands would create. Kariba has, in normal times, provided another routeway across the Zambesi, while the power has vastly reduced Zambian needs of Rhodesian coal and thus has decongested Rhodesian railways by freeing them from most long-hauls of coal to the copper belt. Part of the Katanga line in the Congo (Kinshasa) has been electrified with hydro-electric power, and its capacity thereby increased. In Ghana it was decided to build a second deep-water port at Tema even before a final decision was taken to build the Volta Dam. The port was, in any case, almost essential to the future development of Ghana; it was vital to the construction of the nearby dam.

On the other hand, transport needs or improvements may make possible the construction of a barrage or dam. In Mozambique the Limpopo barrage and multi-racial irrigation scheme had to await Rhodesian need for a railway to Lourenço Marques, as an alternative to Beira, since the Portuguese could not afford to build the barrage alone. As built it carries the railway across the Limpopo.

In turn, improved communications may, if other facilities are offered, encourage the development of commercial fishing and tourism on lakes behind dams, as on Lake Kariba, and as intended on Lake Volta. The latter is also expected to become a new and cheaper transport artery to and from northern Ghana. Lakeside ports are to be provided for this purpose, and the ultimate provision could be a service of truck-ferries with roll-on and roll-off facilities. Akosombo could become an inland port and industrial town, somewhat akin to Jinja in Uganda.

It has been suggested that all dams should be as multi-purpose as possible but that, in any case, their effects are inevitably multi-purpose. Dams for irrigation may encourage land reform, new techniques and new outputs. Dams for power for one main purpose (as in Ghana) may greatly improve the productivity of other industries and mines, facilitate irrigation on the fringes of their lakes or by provision of power for pumps, and provide domestic water supplies.

Having said so much of large and well-known dams one should

not lose sight of the fact that small dams, often of rock or earth, may not get much notice in literature but may be of great local importance, especially where large dams cannot be built, or at least not for many years. Small dams should be encouraged to improve water supplies for man and animals, especially in the really dry and poor countries.

By harnessing water we are taming a natural element, one of man's most fundamental assets, an essential of civilisation, yet one in short supply.

[1] Monica M. Cole, 'The Kariba Project', *Geography,* XLV, 1960 pp. 98–105.

[2] M. A. Selim, 'The High Dam Project', *Bulletin de la Société de Géographie d'Egypte,* Septembre 1955, pp. 113–123.

[3] D. Hillings, 'The Volta River Project', *Geographical Magazine,* XXXVII, 1964–5, pp. 830–41.

[4] On sites and types of dams see J. E. Richey, *Elements of Engineering Geology,* 1962, especially Chapter 7.

[5] W. H. Reeve, 'Progress and Geographical Significance of the Kariba Dam', *Geographical Journal,* CXXVI, 1960, pp. 140–6.

[6] R. J. Harrison Church, *Some geographical aspects of West African development,* 1966, pp. 27–30.

[7] Harrison Church, *op. cit.,* pp. 31–36.

[8] A. M. O'Connor, *An Economic Geography of East Africa,* 1966, pp. 145–7, analyses the reason why this is so.

[9] W. H. Reeve, 'Progress and Geographical Significance of the Kariba Dam', *Geographical Journal,* CXXVI, 1960, pp. 140–6, includes a map of the new settlements; Monica M. Cole, 'The Rhodesian Economy in Transition and the role of Kariba', *Geography,* XL, 1962, pp. 15–46, and *op. cit.,* pp. 102–4.

[10] *Volta Settlement Symposium Papers,* Volta River Authority, Accra, and University of Science and Technology, Kumasi, Ghana.

## BIBLIOGRAPHY

K. M. Barbour, *The Republic of the Sudan,* 1961.

......................., 'Irrigation in the Sudan: its growth, distribution and potential expansion', *Transactions Institute of British Geographers,* 1959, pp. 243–263.

R. J. Harrison Church, 'Observations on Large Scale Irrigation Development in Africa', *Agricultural Economics Bulletin for Africa* (United Nations Economic Commission for Africa and Food and Agriculture Organisation of the United Nations), No. 4, 1963.

......................., *Some geographical aspects of West African development,* 1966.

Monica M. Cole, 'The Kariba Project', *Geography,* XLV, 1960, pp. 98–105.

......................., 'The Rhodesian Economy in Transition and the Role of Kariba', *Geography,* XLVII, 1962, pp. 15–46.

H. E. J. Davies, 'Irrigation Developments in Sudan', *Geography,* XLIII, 1958, pp. 271–3.

A. Gaitskell, *The Gezira,* 1959.

W. A. Hance, 'The Gezira: An Example in Development', *Geographical Review,* 44, 1954, pp. 253–270.

......................., *African Economic Development,* 1958.

D. Hilling, 'The Volta River Project', *Geographical Magazine,* XXXVII, 1964–5, pp. 830–41.

B. S. Hoyle, 'Economic expansion of Jinja, Uganda', *Geographical Review,* 53, 1963, pp. 377–388.

L. Huszar, 'The Volta Resettlement scheme', *Journal of the Town Planning Institute,* 51, 1965, pp. 279–282.

Sir Robert Jackson, 'The Volta River Project', *Progress,* 50, 1964, pp. 146–61.

J. H. G. Lebon, 'On the Human Geography of the Nile Basin', *Geography,* XLV, 1960, pp. 16–27.

D. C. Ledger, 'The Niger Dams Project of Nigeria', *Tidschrift voor Economische en Sociale Geographie,* 54, 1963, pp. 242–7.

A. M. O'Connor, *An Economic Geography of East Africa,* 1966.

W. H. Reeve, 'Progress and geographical significance of the Kariba Dam', *Geographical Journal,* CXXVI, 1960, pp. 140–6.

J. E. Richey, *Elements of Engineering Geology,* 1962, Chapter 7.

M. A. Selim, 'The High Dam Project', *Bulletin de la Société de Géographie d'Egypte,* 1955, pp. 113–123.

*Volta Settlement Symposium Papers,* Volta River Authority, Accra, and University of Science and Technology, Kumasi, Ghana.

G. J. Williams, 'The Guma Valley Scheme, Sierra Leone', *Geography,* L, 1965, pp. 163–6.

CHAPTER II

# What An Economist Wants To Know About Dams

J. L. JOY

THE job of the professional consulting economist is to improve decision making. Only too often the only decision posed to an economist is whether or not to build a dam that has already been sited and designed. In this situation, the economist is usually required to place some sort of value upon the proposed investment—a benefit-cost ratio or a present net worth estimate[1]. If this value is high enough it is concluded that the dam should be built. This is a poor way of making decisions when the alternatives are many and complex and it involves a most inefficient—even dangerous—use of an economist.

What an economist wants to know about dams depends upon the nature of the decision and the way in which he is involved. Typically, with water basin development, there is a high degree of interdependence between all decisions. Here we shall take a broad view of the range of decisions with which an economist might be concerned. In practice, a sequence of decisions has to be made; a sequence unfolding through time starting, perhaps, with a decision to commission a consultant's preliminary report and continuing in perpetuity as successive steps in the development of the water basin. In considering what an economist needs to know about dams I shall argue that an economist needs to be party to the decision-making process from the start and I assume that he is not being simply called in—as he usually is—to provide an evaluation after a scheme has been outlined.

Unfortunately, in reality some of the most important decisions relating to the building of a dam are likely to be taken without effective participation by economists or indeed of others whose views are relevant. Let us consider simply the determination of the site for the building of a dam. Sometimes there may be available only one such site technically determined by the topography of the water basin and by its geology. But, desirable though it may be to build the dam on a firm bedrock, in a form economical of construction materials, at a point where storage capacity per unit cost is maximised, there may be other considerations relevant to the siting of the dam. Where a dam is built will affect its potential as between power generation, flood control and irrigation capacity. It will affect the land that will be irrigated and the land that has to be flooded. It will affect the people who will benefit and the people who will suffer. Insofar as it affects irrigation potential it may have long-term food supply and foreign exchange implications. It may also affect transport and communications—whether or not existing railways and roads become inundated; whether or not water-ways become usable as part of the transport system and at which point communications converge to cross the river. Thus the technically most efficient dam site, narrowly defined, may not be the one most socially desirable and alternative technologies of dam construction and water storage might need to be considered if alternative sites are to be exploited.

An economist who participates fully and from the start in decision making for water basin development needs to know the basic technical strategic alternatives and the range of possible means of water control and utilisation. Economists may become involved with hydrologists in building simulated models of river systems the results of which they may need in order to determine alternative storage and release patterns and their implications for power generation, irrigation, flood control and navigation[2]. Where groundwater is important the alternatives faced may be greatly extended — especially where groundwater in different places is at different depths and of different degrees of salinity and subject to different rates of recharge. In such a case, one might have to choose not only between one or many dams and alternative sites and patterns of operation but also between dams and tubewells, or different mixtures of the two, possibly using dams for power generation for driving tubewells. A further related aspect might be the costs of canal construction or widening and the possibilities afforded of carrying water from dam storage or tubewells to other areas. In all cases, too, the pattern of development through time offers wide ranges of choice. Usually, the choice of the time pattern of development depends not simply upon economic factors, such as the growth of demand for power, but also upon physical factors, such as the rate of silting of a dam or the imminence of water logging with a given rate of rise in the sub-surface water table. Thus an economist may want to know a good deal about the physical aspects of water basin hydrology and engineering.

But in order to identify key alternative development strategies an economist will wish to know a good deal more than what water there is and what can be done with it. His major concern is the prediction of the consequences of alternative patterns of water use. For the purpose of prediction he needs a good deal of data. How much data and what sort depends to some extent upon the objectives in the light of which decision is to be made.

Thus one of the first things that an economist will need to know about a dam is the range of considerations that are going to be operative in the process of decision making. In this respect, the total of additional income generated in the economy by the building of a dam is not necessarily the only relevant consideration. The distribution of this income may be particularly relevant and, related to this, so also might be the number of jobs created. Matters of secondary importance might be the effect of the dam upon food supply, foreign exchange and government tax revenues. Thus prediction may need to be undertaken in respect of all these items—and others too, perhaps.

In the process of eliminating alternatives it will also be necessary to know of policies that government may wish to pursue. The government may not for example be willing to pursue irrigation development on the basis of estates even though these may generate both the maximum increment of national income and the largest number of jobs. Others' objectives besides governments' may need to be taken account of also. If agricultural settlement is to take place on a family farming basis the pattern of farming predicted must be consistent with the objectives of the settlers. An understanding of their norms with respect to such matters as working effort, subsistence production, cash income, sexual division of labour and so on will need to be known.

I have noted earlier that the effects of dams are likely to open up opportunities with respect to irrigation, flood control, power and communications. It would be unnecessarily tedious to itemize the full list of revelant data needed by an economist under each of these headings. But some detailed idea of the range of data requirements might be illustrated in relation to irrigation aspects. Where we are concerned with irrigation potential we will need to know what crops might be grown and the sorts of yields that might be achieved with different irrigation régimes. While this might be relatively simple in the case of a river basin where crops are already grown and where the effect of a dam might be simply to extend the growing season or augment the level of water delivery, it may not be nearly so simple in the case of a river valley where cultivation has not been developed or where it has been extensive dry land farming. With respect to yields, it may be necessary to know both the biological potential of the area and the likely means and deviations of yields which might be experienced by farmers over time. In practice, such knowledge may not exist and effective estimation of these values may require a systematic search for relevant data and its very careful analysis. The resulting 'data' may well be little more than heroic assumptions but even these may be better than unfounded notions. Where water is scarce it might be particularly important to have estimates of yields under sub-optimum watering techniques (i.e. of the extent to which yields fall when water is supplied in some pattern or amount other than that which would give maximum yields). Such estimates may be necessary, in any case, where capital or running costs can be saved by sub-optimal watering régimes.

In appraising major alternative river basin development strategies, one needs to consider for each the best possible pattern of exploitation. That is, if the choice were between one high dam and several small dams it would be necessary to decide what would be the best way to exploit the alternatives before choosing between them. With regard to irrigation developments, where it is decided that the best way to exploit irrigation development is to let farmers develop on the lines that they find most profitable, the economist's role is more that of a predictor than that of a planner. Even so, he may need to plan for the optimum provision of credit, research and extension and marketing services. In some cases, however, irrigation development is likely to be by regulated settlement schemes and here the economist will need to design optimum farming systems which are likely to be followed by settler farmers and to which irrigation design must be related. Provided the design of farming systems is intended to produce cropping and husbandry patterns which farmers will want to follow, there is little difference between planning and prediction. In both cases it will be necessary (i) to calculate the change in potential created by the new or improved irrigation (measured from the farmer's point of view); (ii) to calculate the rate at which farmers will exploit this new potential.

However, both the nature of the potential created and the rate at which farmers can exploit it will be governed by official policy. This might, for example, dictate that settlement will be on family small holdings with little or no hired labour rather than on estates or 150 acre mechanised holdings, or cooperative settlements. Choice of crops might be affected especially by government's willingness to provide processing or market outlets and even by such matters as its readiness to amend regulations governing dates of crop planting and uprooting. In various ways assumptions about government policies may bear critically upon the plans and predictions that the economist is trying to make. In this situation it is desirable—failing clear guidance and an unswerving government policy—to predict or design optimal farming patterns for each set of permuted assumptions about government policy. In these days of electronic computers this is not so impossible as it may sound. The main point to note is that another set of data important to the economist is the set of constraints on government policy which define the context within which planning and prediction must take place.

The problems of defining optimal or expected farming systems may need to be tackled in relation to labour requirements of different crops at different times of the year. Again, where such data does not currently exist, heroic assumptions may be necessary based on the consideration of the most relevant existing data. To some extent it is possible to explore the significance of errors in these assumptions. At least it should be possible to determine which of the assumptions about labour requirements are most critical.

The economist will need to know a good deal about agronomy —about cropping sequences and pests and diseases, for example— and about irrigation technology. Alternative techniques of irrigation may need to be considered. There may be a choice of furrow irrigation and overhead sprinklers. This choice again is not solely a technical matter, though very much technical data will be required relating especially to capital cost implications of alternative systems. (In the case of furrow irrigation the capital requirements of land planing and clearing and, in the case of overhead sprinklers, the capital costs of pumps, pipes and sprinklers and their relation to different forms of control, layout and farming systems). Physical properties of the soil will be important, especially as these relate to e.g. percolation (important with furrow irrigation) and panning. The choice of techniques will affect both capital and running costs in local currency and in foreign exchange. It will affect also the possibilities of obtaining overseas aid; the nature and size of communities that might have to be envisaged and provided for; the efficiency of water and land use; the nature of authority and control of irrigation development as well as those considerations which might ultimately be most important—the number of people settled and the extent of extra incomes created.

In designing — or predicting — alternative irrigation developments a vast amount of technical physical data will be required most of which the economist might ideally expect to have provided for him. There is, however, a good deal of data that he needs that he should properly expect to provide himself. In designing or predicting farming developments he will need to know the prices of factors and products. Where new developments create new demands for factors and increase product supplies, allowance will need to be made for the impact of these changes on prices. In appraising alternatives, opportunity cost prices for factors might need to be calculated where there is reason to suppose that market prices do not reflect true costs.

So far then we have suggested that the economist starts by indicating the alternative strategies of development and their broad implications. Fairly early on a decision will be taken to consider only certain possible lines of development and the economist should play a role in the process of eliminating some alternatives from further consideration. Thereafter, more detailed design or prediction is carried out in respect of the shortlisted strategies still under consideration and their implications for national income, employment and so on are assessed. At some stage decisions will have to be taken about dams—one or many; where located; storage capacity; operation in relation to power generation and irrigation; and the timing of these developments especially in relation to other complementary inputs. Such decisions may not be readily possible on the basis of initial surveys and reports. These might simply reveal the critical nature of some of the assumptions made and the desirability of further investigations even at the cost of delay in going ahead. More information might be required on the hydrology of the river basin, on soils, crop potentials, farming systems, silting rates and the like—even where this means delay in initiating the project. (Not that a decision to build needs to be a decision to build immediately. A dam which is expected to have a short life because of silting might especially be delayed until power demands, for example, have grown to the point where fullest use can be made of the new capacity).

Typically, a detailed survey will lead to some positive decisions —a choice of strategy; a call for design specification or tenders for dam construction; a bid for overseas loans. It will also lead to further investigation of some aspects of development where decisions can be postponed. As we stressed earlier, there is a continuous sequence of decisions to be made and action to be taken by a wide variety of people in government alone, and it is difficult to conceive of any of them where the economist cannot contribute to a clarification of the issues.

In trying to answer the question what an economist wants to know about dams, I have seen the economist's role as having two sides: one, clarification of choices to be made in respect of water development policy; two, the contribution to the design of development. In both these roles the economist will need to have available a vast mass of technological data covering a wide range of disciplines from hydrology to agronomy as well as more strictly economic information relating to such issues as market conditions, expenditure patterns and tax systems. It is clear, however, that decision making based solely on technical and economic considerations is wholly unsatisfactory and that the social and political implications of the choice of alternative strategies of development, including especially alternative technologies of irrigation, has profound social and political implications. In both the design and predictive aspects of the economist's work, he will need to take account of political and social aspects and, in the choice of strategies, the political and social implications of alternative lines of action will need to be clearly presented.

Thus, in all aspects of his work, the economist must clearly operate in the closest cooperation with technologists from many disciplines and with sociologists, political scientists and administrators. To be effective in his own work he needs to know a good deal about theirs.

[1] See S. A. Marglin, *Public Investment Criteria,* London, 1967.

[2] See A. Maass et al., *Design of water resource systems,* Harvard, 1962, and A. S. Manne, *Product-mix alternatives: flood control, electric power and irrigation,* Carter Foundation Paper No. 175, Yale, 1962.

CHAPTER III

# Resettlement

DAVID BROKENSHA

and

THAYER SCUDDER

## INTRODUCTION

THE creation of large man-made lakes in Africa has been responsible for the relocation of large numbers of people, including some 50,000 people displaced by the Kariba Dam, over 70,000 people by the Volta Dam, and over 100,000 by the Aswan High Dam. In Nigeria the lake that will form in 1968 behind Kainji Dam will displace another 50,000 people. With further dams planned for all of Africa's major river systems, this is just a beginning.

Population movements are, of course, nothing new to Africa. In the past, however, they have been primarily 'voluntary' in that people chose to move rather than to remain in a situation which they considered, for various reasons, unfavourable. While this is still the case in contemporary Africa, the proportion of people involved in 'involuntary' movement has been increasing. Such movements are usually the result of government decisions, which the people have no option but to obey. The various reasons for the decisions include national security, threat of epidemic disease, environmental degradation, urban redevelopment and environmental alteration as a result of engineering projects. In all cases, once the policy decision has been made, the people involved have no choice in the matter: the move is compulsory. For that reason alone the government, as the initiating agency, assumes a degree of responsibility quite different than is the case with voluntary movement, or even with refugees, of whom there are some half a million in Africa in the care of the U.N. High Commissioner for Refugees. This is especially so with resettlement in connection with man-made lakes, since these are a product of government attempts to improve the quality of the physical and social environments of their citizens.

Population relocation is a complicated process, which because of population imbalance alone is bound to increase in scale throughout Africa. Yet to date little is known about the response of people to compulsory movement. It is also unfortunate that individuals, agencies and countries that have experience of population resettlement are seldom consulted by those responsible for new projects. While the former gap in our knowledge can only be filled in through careful research, the exchange of information between those involved in population relocation and the use of skilled resettlement administrators can minimise the risk of the same mistakes being made time and again. In the meantime, we hope that this chapter will be useful in placing population relocation in a wider perspective and in showing some of the options, hazards and pitfalls involved. The topic is a complex one and one which warrants much more attention than it has received.

## POPULATION RELOCATION AND RIVER-BASIN DEVELOPMENT

The purpose of development presumably is to make areas more habitable for people. It requires the best use of human resources (surely the most valuable of any developing country) in terms of the individual and his total environment, the latter including not just the physical and biotic environment but also the society to which the individual belongs and the international community of societies that impinge upon his society. To be successful, development must aid the emergence of relatively creative, relatively productive and relatively integrated human populations—populations 'whose potential is productively utilised rather than drained off under conditions of individual anxiety and social stress'[1].

According to the 1963 Report of the United Nations Conference on the Application of Science and Technology for the Benefit of Less Developed Areas, 'one of the greatest dangers in development policy lies in the tendency to give the more material aspects of growth an over-riding disproportionate emphasis. The end may be forgotten in preoccupation with the means. Human rights may be submerged and human beings seen only as instruments of production rather than as free entities for whose welfare and cultural advance the increased production is intended.' This statement is particularly appropriate to African river-basin development projects where international and local planners have tended in the past to over emphasise tangible benefits such as increases in power and gross national product as opposed to the development of the lake basin area and its inhabitants. At Kariba we do not think it is unfair to state that the former Federation government viewed the lake as merely a byproduct of the dam, relatively unimportant in itself, and the people requiring relocation as an expensive nuisance. In Nigeria, international experts consistently belittled the agricultural and fisheries potential of the Kainji Lake Basin population in spite of the fact that over 10,000 of the people were involved in one of the most intensive farming systems in sub-Saharan Africa. These people had highly developed irrigation skills (used for cash-cropping onions) which are as rare in tropical Africa as they are necessary for feeding tomorrow's population. Turning to fisheries, one influential evaluation during the feasibility surveys concluded that the local population had no aptitude for fishing. Again, this statement, which presumably had considerable influence on policy decisions (including such critical matters as bush-clearing) is grossly inaccurate since fishing is a respected part-time activity among a majority of the population and a full-time occupation among a significant minority. Furthermore, preliminary evidence gathered by a team of Nigerian and British biologists in 1964 suggests that the general absence of large adults of commercial species is a result of the effectiveness of local fishing techniques. Such an underestimation of local human resources is most unfortunate and is hardly in the interests of either the lake basin population or Nigeria.

Though it may not be too late to re-evaluate the situation at Kainji on the basis of recent information gathered by Nigerian and FAO experts, it will not be easy to act on this information since funds have already been allocated for different aspects of the Kainji project. As one of us has written elsewhere, 'The logical corrective here is to broaden the feasibility studies, which are apt to stretch out over a number of years prior to site selection. To date, the scope of these has been much too narrow . . . Though expansion of the usual geological, economic and engineering surveys will increase costs, these are slight in terms of the cost in both human and financial terms of a prolonged period of famine relief. They are even smaller when measured against the benefits that could accrue from a well-planned, well-timed and well-implemented programme of river and lake basin development'[2]. We will be referring later to certain aspects of this statement. Suffice it to say now that population relocation presents an exceptional opportunity to improve the standard of living of those involved as well as to contribute to national development. 'With proper timing and planning, new environments with improved social services can be created in carefully selected relocation areas. Through experimentation before and after resettlement, new production techniques can be developed to increase per capita income without depleting soil fertility and local resources'[3].

To date this opportunity has been neglected in Africa. While most river basin development projects have indeed had multi-purpose aims, too often these had not included the people undergoing relocation. In this sense there is still much to learn from the success of the Tennessee Valley Authority. Well-known as a conscious attempt to achieve a multi-purpose development of resources, this project, completed in the 1940s, also benefitted those undergoing relocation. To quote David Lilienthal, then chairman of the TVA, 'The TVA could not close the gates of the dam, pay off the landowners and townspeople, and call it a day. That would not do because the resources of the region—human energies included—were to be seen as a whole . . . What at first seemed a calamity was turned into an opportunity, and a community sense of direction has resulted that continues to bear fruit'[4]. These words, though written over twenty years ago, are very relevant to our discussion of contemporary Africa, yet they need constantly to be repeated for the main points—a total integrated development plan, creating of an opportunity, and the involvement of the people—have still not been fully grasped by all planners. Lilienthal stresses the importance of involving both the local people and the relevant agencies— 'The unified development of resources requires the broadest coalition of effort. This is a job not only for all the people, but for all of the people's institutions . . . The task calls for a partnership of *every* agency of government, state and local as well as federal, that can further the common purpose'[5]. This final point is most significant: an integrated approach implies the use of existing institutions, and the creation of conditions in which cooperation between them is possible. As we shall see, interdepartmental jealousies and lack of administrative liaison can be extremely inhibiting factors in implementing resettlement plans.

## THE PLANNING AND TIMING OF RESETTLEMENT

Planning can be broken down into at least three aspects. There is the development of the physical environment including both the lake basin and the resettlement areas, including an infrastructure of roads, buildings, schools, markets, etc. There is social development including effective utilisation and alteration of existing institutions, development of new institutions where necessary, encouragement of new attitudes and provision of the necessary teachers, medical aides, agricultural extension officers and the like. And there is planning for the actual movement of the people, their livestock and their belongings, and their reception in relocation areas.

Resettlement planning and execution takes *time*. Data must be accumulated on population numbers and structure, on relevant aspects of social organization and values, on resources requiring compensation, and on water and land resources in potential resettlement areas. The administrative structure must be set up and suitable personnel recruited. Policy must be established on government expectations of the local population, and on what the local population can expect from the government in terms of establishing viable villages and economic systems. Effective communication channels must be created to keep the people as informed as possible and to bring them into the planning of resettlement, for after all it is their lives that are involved. Roads must be built, along with schools and other social services; the lakeshore margin demarcated as accurately as the situation requires and personnel and funds allow; an adequate and reliable supply of water provided; bush cleared for village sites, fields and fish grounds; and new villages built and economic systems developed. The actual execution of the resettlement itself requires careful planning if loss of property and livestock is to be minimised and deaths of people are to be avoided. And if gains planned are actually to be achieved, follow-up for a number of years subsequent to relocation is essential.

Up to now the time allocated for planning and executing resettlement has been completely inadequate to undertake these tasks prior to dam completion and lake basin inundation[6]. At Kariba, Volta and Aswan the results have been the same—the people have been relocated before the new environments were prepared to receive them. In what was then Southern and Northern Rhodesia it took the relocated Tonga two years to clear sufficient land to meet their bare subsistence needs; during this period inadequate harvests made them dependent on government organized food relief. In Ghana, one year after the relocation of a majority of the people, the government's ambitious agricultural programme was partially successful in only one of 52 planned communities. Here World Food Programme relief was necessary and was still being given over two years later. World Food Programme relief was also received in the Sudan in connection with the movement of Nubians from the Sudanese portion of the future Lake Nasser to Khasm el Girba. Egypt itself was luckier, not because the land was ready for irrigation in the resettlement areas (in fact less than 10 per cent of the acreage set aside at Kom Ombo had been reclaimed at the time of relocation) but because the Nubians had an alternate source of income through labour migration. In all three cases the initial opportunity to fit the evacuees into a more productive environment has been lost. Instead, the people are faced with a demoralizing transition period that in some cases may last for over a year. During this time period there is a definite risk that the more progressive individuals will seek work elsewhere. Indeed, at Volta this is already the case, with Lawson noting that settlers are leaving certain new towns to farm elsewhere. This was so even at Nkwakubew, the first settlement in which the government agricultural programme got under way. Because they received only about £10 for an acre of maize, instead of the £30 anticipated (and this was less than they earned previously), 'about 20 farmers have now left the settlement town to farm in the Anum area, 20 miles away, where they were able to cultivate up to 10 acres from which in the last season they earned incomes of between £40 and £100 over six months'[7].

These unfortunate situations were in large part a result of the fact that the planning of resettlement was delayed until after the initiation of dam construction. The evidence is conclusive that between then and the completion of each dam there is insufficient time to plan and execute an effective resettlement policy. Instead, resettlement becomes a crash and tension-ridden programme to move the people physically before the lake begins to fill. The logical corrective to this situation is to begin data accumulation and planning early, at the same time as the other pre-investment feasibility surveys.

### THE STRESS OF RESETTLEMENT

Those responsible for the execution of resettlement both in Africa and elsewhere have not paid sufficient attention to the tremendous stress that accompanies relocation. Love of birthplace, no matter how inhospitable it may appear to strangers, quite possibly is a universal human characteristic. To be wrenched from home, as a result of what seems to be an arbitrary and unjustifiable government action, is especially difficult for relatively isolated human populations whose members (and their ancestors) have derived most of their support from local resources for as long as they can remember. In the United States an effective television show has been woven around a TVA drama involving an old matriarch who refused to give up her island home, and the government official seeking her eviction.

Elsewhere in the United States members of a number of Indian tribes have been relocated after dams were constructed on reservation land. The American Indians have good reason to be suspicious of government promises; their attitude is exemplified in a recent popular folk-song by Buffy Sainte-Marie, herself Indian, who sings a harsh and searing indictment of the promises of the white man. The song also conveys the scepticism about the dam and its supposed benefits, an attitude widely shared by people who have to move because of government actions. In January 1967 the last homes of Foster Park, a small mountain pass town in Southern California, were destroyed to make way for a new freeway. One of the 300 white residents complained that they had been promised by the Highway Department that the town would be spared; others said that the freeway route was a mistake, and that they had not received adequate compensation on their property (*Los Angeles Times,* 24 January 1967). The relevance of this experience to Africa is that it indicates that certain attitudes, and causes for resentment, are universal.

Though reliable statistical information is not available, we have reason to believe that morbidity and mortality rates have risen immediately following resettlement in connection with the major African man-made lakes projects. This applies especially to the very young and the very old, with the risks rising where population densities are increased as a result of relocation, where people are relocated in a quite different environment, and where nutritious food is inadequate during and following resettlement. It is the responsibility of government to be aware of these risks and to do everything possible to minimise them. Where the evacuees have been expecting relocation for weeks, months or even years, they may have so reduced agricultural and other economic activities as not to have enough food to sustain them during the actual move and for the days immediately thereafter. Such a situation may even exist prior to the move, especially if the people have been uncertain about evacuation dates or if these have been altered by government decision. Careful planning is necessary here to make sure that the right kinds of food, in sufficient amounts, are available at the proper time so that stress and uncertainty are not increased by diet deficiencies, either in quantity or in quality. It is far preferable to move people from one productive situation to another with minimum food relief.

We do not mean to imply that all segments of a population will actively oppose relocation: on the contrary, we have examples where relocation has been welcomed. But even in such cases, those involved still have doubts about government intentions, which can be reduced if government officials are both aware of the stress involved and make every effort to create effective lines of communication with all segments of the population. With the very old and the more conservative this will be hard—in some cases perhaps impossible—but the attempt must be made. Moreover, every effort should be made to reduce irritants associated with the move itself. While a few broken pots and some dead goats may appear to be incidental, and in fact almost inevitable, to relocation officials, they may be interpreted quite differently by the evacuees[8].

Stress also occurs when people are taken to view and select potential relocation areas before any attempt has been made to service them with roads, schools and other facilities. At the time of viewing, such areas may be completely unoccupied, often for good reasons (such as lack of water or tsetse fly infestation, of which the people are well aware). Though governmental officials may attempt to describe how such areas will eventually look, it is one thing to hear words and quite another to believe them. Will, indeed, the government be able to provide a suitable water supply and control tsetse fly populations? Though it is quite impossible to eliminate such fears, we believe they can be significantly alleviated if the resettlement authority makes an early attempt to provide within a pilot relocation area roads, a reliable source of water and the type of social services that the people themselves desire (and in contemporary Africa there is bound to be some gap here between government policy and popular expectations). Where major changes in agriculture are anticipated, or where new soil types are to be used, pilot demonstration farms should be established as soon as possible. Not only can these be used to show to the evacuees at the time of relocation the type of innovations that the government has in mind, but they can also serve as experimental farms to indicate whether or not the new farming system will in fact produce according to government expectations. If housing is to be the responsibility of government, here too pilot dwellings should be built during the planning stage and the people's comments sought. This was done on a very limited scale in connection with the relocation of the Egyptian

Nubian population. Though some comments were received which resulted in certain modifications, the nature of Nubian housing complaints following relocation indicate that both the government and the people would have profited from a more extensive exchange of information about housing proposals.

Perhaps government officials might do well to adopt some of the sales methods of commercial organizations, at least in setting up model homes at convenient points, and zealously encouraging the people (including the women) to inspect them thoroughly.

## GOVERNMENT POLICY ON RESETTLEMENT

A feature of most resettlement programmes is an authoritarianism on the part of the relevant government agencies, coupled not infrequently with a pained surprise at the reluctance of people to accept enthusiastically carefully considered plans. Describing the situation in Egypt, Horton has written 'For a period of some two years before the first and northernmost *omdiyah* was moved, the governor (of Aswan Province) and the Joint Committee (for Nubian Migration) held monthly meetings with delegations of Nubians in an effort to answer questions and to calm fears. There was no Nubian participation in the planning process; the plan had already been worked out in considerable detail by the ministries, and the meetings were held in order to prepare Nubians for moving and to accommodate some of their desires if this were possible within the general framework of the plan. The basic function envisaged for the Nubians was that of cooperation, without which the job of the Joint Committee would be immeasurably more difficult'[9].

This statement refers equally well to the situation at Kariba and Volta. Though in both cases villages were allowed to pick their relocation sites, provided these met with the minimal government criteria for suitability, they had no voice in plan formulation. While we believe that this is undesirable, we are fully aware of the difficulties of bringing local populations into the planning process, especially when relocation is inevitable, as certain policy objectives must be met and government actions must conform to a rigid time schedule. Nonetheless, we believe that more consultation with local leaders and delegations is desirable in formulating plans which are bound to have a profound impact on people's lives. This is especially true where education and other experiences are broadening the outlook and understanding of the people involved.

The extremes in policy goals can be illustrated by the actions of the Southern Rhodesian, Egyptian, and the Sudanese governments in resettling people in connection with the Kariba and Aswan High Dam Schemes. In the former case the government made no major attempt to transform the settlement pattern and the agricultural economy of those involved. Though a few new schools and at least two hospitals were established the general intent was to leave the people in such a way that they would be neither worse nor better off. At government expense they and their belongings were physically moved to the new locales, payment of poll taxes were suspended for two years, and cut and bundled grass for thatching huts was provided along with foodstuffs to tide the people over until once again their subsistence economy could meet family needs. No compensation was given and no attempt made to introduce major innovations; rather the people were expected to build new houses and clear new gardens as in the past. By contrast the Egyptian and Sudanese governments considered the relocation of their Nubian populations as an opportunity to transform the social and economic organization of those involved. This was to be done through government construction in each country of modern irrigation systems for the production of cash and subsistence crops and of planned communities with extensive social services. The contrast between these two policies is reflected in the significant difference between the funds expended per capita for resettlement. In the Southern Rhodesian case these were approximately £50 per capita as opposed to over £250 per capita expended by the Egyptian Government. In the Sudan the cost exceeded £500 per capita if the expenditure on the Khasm el Girba Dam is included.

Such extremes in policy are a reflection of the political ideology and development policy of the governments involved. These of course vary from country to country. In all cases we believe that more attention should have been paid to the needs and expectations of those undergoing relocation. Furthermore, regardless of whether or not the resettlement authority is able to build these into its plans, policy should have been more thought out in terms of the people's strengths and weaknesses, and of the natural resources of the lake basin area. Strengths can be built into development plans; weaknesses hopefully offset by provision of whatever educational, extension, credit, marketing and other facilities are considered necessary. Just as engineers do not attempt to build a dam before they have a detailed knowledge of the river régime and of the river bed and side walls at the dam site, so the manipulation of human populations should not be attempted until the relevant aspects of their present way of life have been assessed.

It is also important to realise that each population's way of life is closely interrelated with their physical and biotic surroundings. For this reason alone, the physical relocation of thousands of people and the transformation of a riverine habitat to a lacustrine one irreparably alters their ecological balance and their social network. We do not mean to imply here that this is either desirable or undesirable, only that it occurs. Hence, though the Southern Rhodesian government may claim that they have not drastically altered the lives of their Kariba evacuees, in fact this is not the case. For one thing, they are now separated from their relatives in what is now Zambia by a major lake and many miles of bush. This in itself has required major changes in patterns of visitation and has had a profound impact on many aspects of the social and economic organization.

In Ghana, an early official guide-line (as often stated by ex-President Nkrumah) was that 'no one should be worse off after the building of the Volta Dam than they were before'. This rather negative statement is in fact deceptively simple. How does one measure what 'not worse off' means? If a man is moved from a ramshackle collection of mud and pole-huts, with each adult having his own bedroom, to a one-room concrete house with room for additions, is he better off? If a fisherman, accustomed to a 50-yard walk from his hut to his dug-out canoe, is moved to an area two miles from the lake, do the greater fishing opportunities make him 'not worse off'?

A more positive approach to the Volta resettlement was adopted by the Working Party appointed in May 1962. Its guiding principles were

(1) To use resettlement as an opportunity to enhance the social, economic and physical conditions of the people;
(2) To improve their system of agriculture to enable them to effect the transition from subsistence to cash economy;
(3) To plan and locate the settlements in a rational manner so that the flood victims as well as others in the Volta Basin area can derive maximum benefits from the changes involved[10].

Even where government policy is designed substantially to increase per capita income, it still is not easy to establish what is in the short and long term interests of relocated populations.

Looking to the future, most, if not all African governments, will probably attempt to use population relocation to change certain aspects of the people's lives. If policy goals are in fact to be achieved, their execution requires that careful attention be paid to a wide range of factors. In the sections that follow we list and briefly discuss what we consider to be some of the most important of these. Our coverage is not designed to be inclusive, for resettlement is a complicated affair. Though some problems are bound to characterise all resettlement schemes, others, no less important, will arise because of the unique set of conditions that characterise each particular situation. They can be anticipated and dealt with only if the experts have an adequate understanding of the people involved and of their surroundings.

### THE FINANCING OF RESETTLEMENT

Resettlement *is* expensive, indeed considerably more expensive than planners have estimated in the past. This is especially the case where governments commit themselves to ambitious programmes where £150 per capita can easily be spent for new towns and intensive dry farming, and double that amount for new towns and irrigation. Though the final accounting has yet to be made in Ghana, the cost of the Volta resettlement programme will probably exceed £G12m., or approximately three times the original estimates (the rise in costs over the ten years between the original estimates and the completion of resettlement accounts for only a fraction of this rise in expenditure). At Kainji the amount originally budgeted for resettlement was also totally inadequate. Though we do not know the extent to which budgets were revised upward in connection with the resettlement at Aswan in Egypt and the Sudan, the figures we have for Egypt are £E13.5m. and for the Sudan £S24m., which includes the cost of a dam across the Atbarra river. If one adds in the Ghanaian case the cost of food relief over at least two years and the loss of human productivity and motivation that is correlated with a relief programme, the amounts involved are even more substantial. This is the point we wish to make. If planners realise from the beginning the high costs of resettlement and start with relatively realistic figures, then perhaps resettlement will be taken more seriously, with greater effort made to ensure that the money is well spent for the benefit of both the local population and the national economy.

### THE RECRUITMENT OF PERSONNEL FOR RESETTLEMENT PLANNING AND EXECUTION

To date resettlement in connection with African man-made lakes has been planned and executed almost entirely by local personnel. This we believe is unfortunate. While both of us have been impressed with the ability and dedication of many of the government officials involved in population relocation, the fact remains that they have had virtually no access to international capital and expertise. This is in complete contrast to the planning and execution of the very projects that cause population relocation. Their feasibility is carefully assessed by reputable firms of consulting engineers as well as by international financing agencies. Their construction is handled by leading civil contractors. Yet prior international cooperation does not occur with the resettlement aspects. There the formulation and execution of policy is entirely in the hands of government employees, most of whom have had no previous knowledge of, or experience with, resettlement. In recruitment as in financing we believe that a more responsible and sophisticated approach to population relocation is essential if the opportunities involved are to be realised.

### ADMINISTRATION AND COORDINATION

The success of any resettlement depends to a great extent on the type and quality of the administration: this is true of development projects generally, and is recognised by the increasing attention being paid in Africa to improving standards of public administration. Although the resettlement administration will obviously be affected by the prevailing standards of national administration, we are here concerned particularly with the organization and responsibilities of the resettlement authority, and its relationship to other interested agencies. What is the composition of this body? To whom does it answer? Are those involved responsible for both policy formulation and the execution of resettlement, or only for the latter? What relationships are maintained with other government bodies? Because of the different traditions and conditions in different African countries there is no single way of administration that can be recommended. What is important is that the standard rules of good administration be followed. There should be adequate staff, who are well trained and who have good working conditions. They should also have clear roles, and adequate financial support. There should be continuity of staff, as a constant change-over is detrimental to morale and efficiency. The administration should also be free of continual political interference; surprisingly, this was the case in Ghana even though at the time ex-President Nkrumah was enjoying the last years of his power, and it was unusual for him to allow any public agency to manage its own affairs, particularly when the Volta Project was so dear to his heart.

As well as the vertical line of authority, from Parliament or President down to the field official, there is the horizontal distribution of power, that is the cooperation that is necessary with other government departments. Friction and jealousies and inertia can seriously delay implementation of decisions, especially when there are several different branches of government involved. The settlement authority will certainly have to work with departments of agriculture (and other natural resources) as well as with such departments as economic planning, health, education, community development and the ubiquitous treasury. It is essential to devise some lubricating machinery to permit the settlement officials to move quickly, for many national administrations have a built-in protective device of delaying action in the interests of caution and financial control. This means that there must be a senior official with power to cut through the red tape, such as the chief engineer was able to do at Volta by reporting directly to the President of Ghana.

The Resettlement Authority should also have an effective consultative machinery in the form of an advisory body, which, like the authority itself, needs to be established as early as possible if it is to be effective. There are at least two reasons for this: first, there will be so much to do that an early start must be made; and second, this will ensure them a more favoured place in the bureaucratic structure than if they were appointed late, when the relationships and lines of communication have hardened. The advisory body should have representatives of both government and non-government agencies, including development banks, universities, and representatives of the people themselves. There may well be among the people to be resettled a university graduate who has studied development problems; even if this is not the case, representation of the people is highly desirable.

We realise that the participation of the people might be difficult to arrange. At Volta, the chiefs (particularly the minor ones who felt most threatened) who might have been expected to represent their people, seemed to be more concerned about maintaining their relative status in the new circumstances than in encouraging any new developments. Still it should be possible, among 50,000 people, to find a few who would make good representatives.

With the assistance of the Advisory Body, the Resettlement Authority should also be willing and able to seek international advice and assistance in connection with the various stages of the resettlement process. International skills can be especially useful in ensuring that timely and well-conceived surveys are carried out, these being essential for effective policy planning.

We also urge that the administration of resettlement be seen as a continuous process, lasting for five to ten years after evacuation rather than as a temporary crash-programme. It is only by continued planning and evaluation that full benefits can be realised[11].

Although there is no generalised model of an administrative organization which can be applied to resettlement throughout Africa, much can be learned from how Egypt handled the resettlement of its Nubian population in connection with the Aswan High Dam. Here 'over-all responsibility was given to the Governor of Aswan Province. This disposition of responsibility almost automatically put Nubian resettlement into a larger context, which is the decentralised and ambitious programme of economic development of Aswan Province'[12]. The supervision of resettlement fell to the Joint Committee for Nubian Migration. Under the chairmanship of an Assistant Undersecretary of the Ministry of Social Affairs, this committee was composed of members representing the various ministries with resettlement and development responsibilities.

Needless to say, a good administrative and advisory organization alone is hardly sufficient. Also needed is a realistic policy, sufficient time to execute this policy, and adequate funds and personnel.

### CONSULTATION

If the communication between the government and the people is faulty, the success of the whole project will be jeopardised; therefore it is necessary for the resettlement authority to take care to establish satisfactory channels of communication. This necessitates qualified staff who are prepared to spend arduous months in the field, patiently discussing resettlement problems. The task of the settlement official at this early stage is to consult with the people, to inform them fully of government plans in regard to the dam, to show where the water will rise to, and to discuss logistics of the move, compensation, provision of new land for settlement, housing and other arrangements. Initially, government officials will meet almost invariably with hostility, suspicion or disbelief. People whose ancestors have lived for generations alongside a river are usually well acquainted with its régime, and may well be sceptical when told that a dam can be built across it and, if built—often many miles downstream—that it will affect them. This has happened at all the dams. The hostility or suspicion arises from the credibility gap that so often plagues governments, colonial or independent. Even in industrialised countries people fear governments, thinking that they are being moved so that their land can be used for other ethnic groups, or plantations, or agricultural projects.While it is difficult to counter such fears, it is essential to persist with a patient and continuous dialogue with the people, explaining the plans of government and encouraging them to participate, as much as circumstances allow, in decision-making. Sometimes too much effort is expended in the preliminary stages in telling the people what a glorious work the dam will be, and how the hydro-electric power will aid industrialisation and progress. It would usually be better simply to announce that the dam will be built, explain briefly why the particular site was chosen, then move on briskly to more positive thinking about the future.

The form that consultation takes will be determined in part by the type of indigenous political organization. At Kainji, for example, the people are accustomed to a centralised form of government, with the powerful *emirs* receiving both extravagant courtesies and close attention to their commands. The *emirs* appoint village headmen who exercise considerable power at the local level, collecting taxes, settling minor disputes, acting as the channel for the *emir's* commands, and for petitions from the people to him. Though the power of the *emirs* and their assistants nowadays has been challenged by new forces, the traditional rulers have maintained sufficient of their power as to be still reckoned with as prominent figures. In this situation it is essential to work with the *emir* and to obtain his permission to consult at lower levels. Otherwise an extra and possibly insuperable barrier would be added to the difficulty of communication.

By contrast, most of the people in Ghana had a much livelier tradition of participation in government at the local level. Though there were chiefs, no chief could simply tell his people what to do, but rather was customarily expected to engage in consultation with them, even though this frequently entailed interminable discussion. In such a situation much more patience is required of the resettlement officials, who must be prepared to talk with small and large groups of villagers on endless occasions, about the same topics. To do this, a large staff is necessary and a few senior men with diplomatic skill and rare personalities. Although seemingly little headway is made, the attempts at communication are appreciated and the move and resettlement are likely to proceed more smoothly than they would in the absence of communication. In Ghana, as elsewhere, people intensely resent being 'treated like chickens', just picked up and moved: patient consultation, particularly over a vital matter like the selection of a new site for their home, is a recognition of their humanity.

In stressing the importance of good communications between people and government, we have been talking of the necessity of providing opportunities for discussions. Because of the urgency of the task, and because of the limited time available, talking could be supplemented by audio-visual aids. Full use should be made of radio or television facilities, where these exist. Films can be made of new sites and new housing types, and shown to the people to augment personal visits. Pamphlets are an effective means of communication, as at Tema in Ghana where the government printed thousands of copies of a 'question and answer' type pamphlet, which dealt with the queries commonly raised by the people in regard to their impending resettlement. Useful as these sorts of aids are, they must be used strictly as supplementary devices to the regular and continuous visits of settlement officials. To indicate the very wide range of matters on which consultation is necessary, reference is made to a list of 'Questions asked by the Villagers' at the Volta Resettlement Project[13].

## SURVEYS

A wide range of surveys and baseline studies are necessary to facilitate the resettlement process and the planning of new environments. Especially important are hydrological, demographic, public health, livestock, social and ecological surveys. If corners are cut in executing these or if they are completed too late to influence planning, the whole resettlement and development policy is endangered. In such a situation there are no gainers, only losers.

Hydrological surveys are of course crucial to ascertain the water resources of relocation areas and the best means to tap these resources. They need be carefully timed, with ample allowance made, for example, for borehole failure (at Kariba, on the southern Rhodesian side, of 233 boreholes drilled only 143 were considered successful by government standards), so that the people can be assured a reliable water supply at the time of relocation. To date this has not been the case, with all the major resettlement projects plagued to a varying degree by water failure. In some cases the subsequent installation of a piped water supply at considerable expense has been necessary.

Unfortunately, accurate information on population numbers and composition is frequently not available in African countries, especially in regard to areas which until recently have been relatively isolated. Ghana, for example, was supposed to have excellent demographic information in comparison to other developing countries. In its 1956 Report, the Preparatory Commission estimated that some 62,500 people would have to be relocated in connection with the Volta Dam. On the basis of the 1960 census, which was regarded as reasonably accurate, the figure was revised to 50,000 people. In 1962 a detailed social survey was made which placed the population at 78,285. Subsequently this figure was raised to 82,000. Ghana's experience has been shared by other countries, none of which has had a reliable population estimate for the areas to be inundated. Obviously an accurate figure is a pre-requisite for any planning, essential alike for financial estimates and for relocation plans.

The purpose of public health and medical surveys is to determine the state of health of the local population prior to resettlement, and to identify ahead of time medical (including nutritional) problems that may arise as a result of relocation. Because of their isolation and the scarcity of clinics and dispensaries, not to mention hospitals, the state of health of many populations undergoing relocation may be worse than elsewhere in the countries involved. According to the health authorities, the Egyptian Nubian population was malnourished prior to relocation. At Kainji, a relocation survey by the Northern Nigerian Ministry of Health of three-quarters of Old Bussa's population of 2,630 showed the calculated infant mortality rate to be 35 per cent. Tests for specific diseases (the sample ranged between 100 and; 300 individuals) revealed that 32·7 per cent had malarial parasites, 5·4 per cent Ancylostoma, 3·6 per cent Ascaris, 12·3 per cent schistosomiasis and 40 per cent had positive skin snips for onchocerciasis. As we have already mentioned, morbidity and mortality rates can be expected to increase immediately following resettlement, for a number of reasons, including the fact that the people are disoriented and under stress. Possible hazards should be identified ahead of time, and means designed to reduce them, including the provision of adequate food during the relocation period. Where populations are aggregated as a result of resettlement (and some increase in density has occurred with all the major resettlement projects) there is an increased risk, for example, of epidemics of bacillary dysentery, measles and chicken pox. There is also the risk that the incidence of tuberculosis may go up along with parasite loads. All such possibilities should be taken into consideration.

In Ghana the Preparatory Commission stated in 1956 that 'the creation of the lake would clearly introduce a number of new and uncertain factors affecting the health of the people'. Dr. Warmann, the chief medical officer of the VRA, has outlined the difficulties of the public health programme, and also proposed ways of dealing with them. He points to the problem of ignorance and apathy, and the difficulties of education in health matters; he urges that health surveys should be conducted before and after resettlement; he considers the relation to health of a good water supply, town planning, diet and agricultural changes, and he stresses that 'resettlement must be accompanied by a corresponding rise in the economic state of the people if they are to cope with and take advantage of the better conditions offered by it'[14]. Once again the central importance of economic improvement is stressed, as is the inter-relationship of all aspects of resettlement; health can only really be improved if there is a rise in the standard of living. The cyclical effect of poverty—poor housing, bad health, low wages, unemployment, low skills, inadequate education—applies in rural resettlement projects as well as in urban renewal.

The numbers of livestock in African villages, and especially of goats and fowl, are usually far greater than outside observers would expect. A relatively accurate and timely census by species is important, so that the resettlement authority can plan realistically for their future. If the decision is made that many must be disposed of before relocation, then some government attempt should be made to ensure that the people get a fair price. Otherwise the sudden appearance of large numbers of animals on the market will depress prices, as occurred in Egypt in connection with the Aswan relocation, and cause hardship to the people involved. On the other hand, if the people are to be allowed to take their livestock with them, then planning for their accommodation, feeding and protection (both from disease and predators) in relocation areas is important. To date such planning has been inadequate. At Kariba, Tonga evacuees in Southern Rhodesia complained that they lost large numbers of stock because of animal trypanosomiasis. In Egypt large numbers of Nubian livestock died in quarantine in Aswan, while a significant proportion of the survivors died in Kom Ombo because of the non-availability of fodder. Though the government had warned the people that this would be the case for at least the year following relocation, nonetheless the people still brought their animals, in part because prices were so low in Nubia, in part because the animals were closely identified with household interests. At Volta no accommodation was provided for the 42,000 sheep and goats and other livestock that the people brought to the new settlements, primarily because the census returns were not in at the time of settlement design. Such situations are unfortunate: 'Inevitably they make those involved unhappy; at worst they increase the settler's resentment against relocation and government and create an environment unfavorable to cooperation'[15].

Although social surveys are often conducted together with the human and animal census, they are considered separately as they do have rather different purposes. Varied use has been made of social surveys. At Kariba, there was no systematic attempt to gather a wide range of information on the people. In Ghana the Volta Resettlement Authority probably made the greatest use of surveys, using well-designed questionnaire forms in July–October, 1962, and analysing the information obtained to assist in planning new settlements. At Aswan, forty social workers conducted social surveys over a thirty-day period, but this not until 1960, so that, as at Volta, the results were too late in terms of timely planning.

The purpose of social surveys is to provide information on the people undergoing relocation that will assist planners to formulate and execute policy. If this information is to be effectively used, it should be gathered as part of the pre-investment feasibility surveys. Experts should be recruited to formulate and phrase whatever questionnaires are used, and to process the information gathered into a form comprehensible and useful to the planners[16]. Wherever possible they should be citizens of the countries involved, although it is important they be willing and encouraged to draw on international experience and expertise where necessary. Also it is important that they have a clear idea of just why their services are required, so as to minimise the risk of too academic an approach to the problem at hand and delayed processing of topical information. According to E. A. K. Kalitsi, former head of the Volta Resettlement Department, 'if only the sociologists involved in this type of work could try to meet some of the day-to-day requirements of the job, their value to the administration would be much greater'[17].

The nature of social surveys will, of course, vary from situation to situation. Where government has taken over the responsibility for building acceptable new settlements, information is needed on present settlement patterns, house types and social networks between houses. In connection with Aswan Resettlement, 'Four sizes of new houses were built, and for ease of construction houses of the same size were grouped together. The grouping of families by size not only broke up the old neighbourhoods and villages within each district, but also segregated most of the older members of the community. Widows or elderly couples whose children had their own house were assigned to the small-house section of the new community. Thus it is often difficult for their younger relatives, who live in the section of larger homes, to render the assistance customarily due elders'[18]. This situation is still a cause of complaint and hardship and indicates only one area in which a well-designed social survey can be expected to provide information that could assist planners in reaching a more equitable solution to the many complex problems that resettlement presents.

Where new systems of farming are planned, survey information on the present system of land use as well as on the attitudes of the people toward farming and toward past attempts by government to introduce agricultural innovations is important. Information could also be gathered on the nature of leadership (as an aid, for example, to forming local councils and achieving community development goals) as well as on other relevant aspects of the social organization and values of the peoples concerned. The above are only a few examples of the type of information that would help planners design new environments which are both acceptable and more productive. Surveys can also provide information which will facilitate communication between planners and people by providing the former with some insight into the behavioural patterns and values of the latter. In this way the expectations of the planners will be more reasonable and realistic, and they should not be disconcerted by the failure of the people to respond to what they consider an entirely rational approach.

Some surveys have included provision for an investigation of attitudes of the local people on a number of matters connected to resettlement. Farmers have been asked where they would like to have their new farms, families have been questioned about their preference in new houses, their neighbours or the size of new settlements. There are dangers in this approach, for it may serve only to raise hopes that are bound to be disappointed; for decisions must eventually be made in accordance with some sort of overall plan and not on the basis of the sum of individual opinions. It would perhaps be better to decide first what range of possibilities are open to government, for financial, administrative and developmental reasons. The people then can be asked to indicate their preference when officials say, for example, 'it is possible for you to go to area A or area B; which would you prefer, and why?' Or officials can ask which of three or four proposed house designs appeal, choosing as examples those which are financially and technically practicable. And in any attitude survey, particularly if it is concerned with household or domestic matters, the woman must be included. This may seem too obvious a point to mention, but experience of many surveys has shown that there is a frequent omission of half the population from consideration, with later disappointing results if the male attitude is presumed to be the prevailing one.

Under ecological and economic surveys we have in mind those relating to the development of new economic systems in the resettlement areas. Information is needed on the present economies of the people to be relocated, on the riverine fish population and on the soil and grazing resources of potential relocation areas. Such information obviously is essential if policy is to reflect government goals, the abilities and interests of the local population, and the nature of available resources. Under land use surveys topics of importance include techniques (including those of a magical or ritual nature) and implements used, the length of the agricultural season or seasons, the division of labour by sex and the amount of time spent by each type of labourer for different crops and activities throughout the agricultural cycle. Such information is relevant for an appraisal of the strengths and weaknesses of the people to be resettled in terms of whatever agricultural innovations the government may wish to introduce following resettlement. Strengths can be built into more highly productive systems, while the development plan itself and the nature of the extension and community development services, of new techniques and implements, and of credit and marketing facilities, can be designed to offset weaknesses.

Far too often the necessary surveys, if initiated at all, are begun only after dam construction has started, so that the relevant information is not available in time for planning purposes. Because at Kainji systematic information on the substantial lake fisheries potential was not available during the planning stage, insufficient funds for bush clearing were allocated. Yet if the necessary funds are not made available, and 'if selective bush-clearing is not carried out prior to dam completion in 1968, commercial fisheries may never develop which would support the uninformed projections of the original planners'[19]. This would be most unfortunate since the full development of the Kainji fisheries could probably support up to 2,500 fishermen plus their dependents at a standard of living equivalent to industrial workers (and few Nigerian industrial plants can be expected to employ a similar number of people on a continuous basis). This is only one example of many that we could give. As noted earlier, the only corrective is to initiate the necessary human and natural resource surveys prior to dam construction—that is, during the period of the pre-investment feasibility surveys which are apt to cover many years.

### LAKE SHORE MARGIN DEMARCATION

Accurate demarcation of the high water contour of the lake shore margin is an expensive and time consuming operation. At Volta the cost was £72 per mile, and at one time 'Practically the whole staff of the Ghana Survey Division was . . . turned on the job. Even so, other surveyors had to be hired . . . . By the end of 1963, 1,940 miles had been demarcated and a decision was taken to curtail the work. By that time the areas with the highest density of population had been covered, and it was decided to take a calculated risk on the balance'[20].

The need for accurate demarcation varies according to a wide range of circumstances, including the nature of the topography, the accuracy of existing maps, compensation policies, population distribution and location of resettlement areas. At Aswan demarcation was not necessary in connection with relocation since all resources to be compensated fell well within the lake basin area and since the people were not to be relocated around the lake edge. At Kainji, where current policy is to resettle the people wherever possible at the edge of the lake, careful demarcation was carried out. Here the gradient of the lake shore margin area was shallow, so that it was not clear whether certain villages would be flooded or whether certain potential resettlement areas would in fact be above the high water line. Furthermore, since the farmers had gardens both within and outside of the future lake basin, an accurate demarcation was necessary to determine which gardens need be measured for compensation purposes.

Under most circumstances, some form of lake shore margin demarcation is important for resettlement and development purposes. The actual circumstances will vary from case to case and need be carefully evaluated in the light of local conditions and government policy. At Kariba no accurate demarcation was made.

Though the need was less than elsewhere, because of the absence of planned communities and the nature of the agricultural system, unnecessary suffering occurred in those villages which with government approval had unknowingly been rebuilt below the high water mark. Such poor planning not only increases the stress of resettlement, but also adversely influences the people's attitudes towards any government attempts to alter their lives.

### COMPENSATION

The inundation of the area to be flooded necessarily entails the loss of a series of individual and community rights. This requires government to establish as soon as possible a consistent and meaningful policy on the type (if any), the amount, and the timing of compensation. All states have the power to acquire lands when this is held to be in the public interest; what varies is the method of acquisition, and especially the valuation, for compensation purposes, of rights in land (cultivation, hunting, gathering, etc.), crops (especially tree and other perennial crops), buildings, shrines and cemeteries, clay and other deposits, etc.

To date in Africa there has been a considerable range in compensation policy in connection with man-made lakes. At Kariba the people were transported to new areas where they were expected to build new houses and clear new gardens with minimal governmental assistance. The Southern Rhodesian Tonga received no compensation at all except in the negative sense of being exempt from poll tax for two years following relocation. In Northern Rhodesia the authorities paid out close to £1 million for approximately 30,000 evacuees. Over half of this was given as a lump sum to the Local Authority, with the rest going to the people. Though some household heads subsequently spent most if not all of their compensation funds for grain during the hunger period that followed relocation (unlike the Southern Rhodesian authorities, Northern Rhodesia did not provide free grain during the two-year period that it took the Tonga to regain complete self-sufficiency), others were able to use their cash payments for the purchase of ploughs, oxen and other productive goods. At Aswan, the Egyptian government surveyed the people's agricultural resources (here consisting of land and tree crops), paying out compensation on the basis of the survey results. According to Fernea and Kennedy[21], the approximate range of compensation per household was from £E40 to over £E1,500. This did not include housing; in both Egypt and the Sudan evacuees received improved housing at government expense. Both governments also intended to provide irrigated land for each household. In the Sudan this would be in the form of both tenancies and free holdings. In Egypt the original intention was for the Nubians to pay for the land over an extended period, although in fact the government may not press for payments[22].

In Ghana compensation was given for crops rather than land, with the VRA drawing up a comprehensive schedule that gave the valuation of all possible crops according to their age, size and quality. New land, cleared and provided at government expense, was to be issued in the resettlement areas. As for housing, government gave the settlers the option of accepting cash, the valuation being based on replacement value less depreciation, or receiving a core house in the appropriate relocation area. Most of the settlers chose the core house, which cost government approximately £250, much more than the assessed valuation of nearly all of the people's own houses.

This brief description indicates the wide range of compensation policy in connection with three major man-made lakes. The important thing is that whatever policy is followed be carefully thought out in terms of governmental goals and the needs and legitimate expectations of the population to be relocated. Ideally policy should be clear from an early stage, since vacillation on the part of government can easily increase discontent and may encourage extravagant and unrealistic claims. At the same time it should be sufficiently flexible to deal with instances of individual hardship. Some people will suffer more inconvenience than others from relocation, and hence deserve special consideration. This is especially true of specialists whose skills may no longer be needed in the relocation areas. An example are the Nubian merchants who serviced the old Nubian communities along the Nile from their *falluka* sailing boats. Yet in the resettlement area where the people are aggregated a few kilometres from the river their services are no longer needed. Where cash compensation is involved for special skills and other resources, including land and houses, decisions must be made as to whether or not payments should be staggered or given before, during or after resettlement. In some situations early payment may have a beneficial effect; in others they may encourage the settler to minimise preparations for the impending move.

There is also the question of compensation for certain 'public' goods, including schools, burial grounds, churches, mosques, and shrines. In Ghana government accepted in *principle* the people's right to receive compensation in respect of certain shrines. Godfrey Amarteifio, the principal resettlement officer, defended the special grant of £10,000 that was made available for propitiations and libations to the ancestors and the gods. 'Libation is an age-long institution in Ghana and is the very basis upon which religious organizations of most communities in Ghana rest. The institution of libations affords a means whereby contact with the supernatural world is established and the ancestral spirits and gods comunicated with.' The VRA, in their social survey, 'recorded and graded fetishes and gods as personal, family, tribal or communal, and further into moveable or immoveable'[23]. The expenditure of time and money was considered justified by the resultant goodwill. Similar policies toward compensation for shrines have also been adopted elsewhere.

### LIAISON WITH HOSTS

Some governments have been able to find 'unoccupied' land—unoccupied in the sense not only that there has been no-one in physical residence on the land, but also that no group has been proved to have valid rights over the land. But in many cases, suitable agricultural land is so scarce that the settlers have had to move in with alien groups, or on land claimed by others. In such cases, friction can easily occur, arising from the resentment felt by the owners of the land towards the newcomers who have been foisted on them, and from the anger of those who have had to leave their old lands to live with strangers. Such feelings are inevitable, but judicious actions and explanations by resettlement officials can help to cool tempers and to promote cooperation. For example, the original farmers often envy the newcomers for being provided with so much, ignoring what losses they have incurred. In areas of Volta resettlement where there are host farmers, they have been explicitly invited to participate in some of the new developments, with officials pointing out to them the advantages of having a new school, or market, or road—as a result of the new settlers. At Kariba Chief Sikongo of the Lusito area of what was then Northern Rhodesia not only allowed some 6,000 evacuees to be resettled in his chieftaincy, but also agreed that they could remain under the jurisdiction of their own chief. The benefits that he has received in return are hardly equivalent, although his people have been able to profit from new schools and medical facilities.

NEW SITES

We now come to consider a part of the planning that is, with the agricultural planning, the most important ingredient of resettlement: we refer to the decisions that have to be made concerning the location and type of the new communities, their spatial arrangements in relation to each other, and the provision of an infrastructure for social and economic development. The aim is to create conditions for the rapid establishment of viable and prosperous local communities, capable of directing their own further growth within a relatively short time. So we examine the problems of town-planning, housing, public services such as roads and markets, social services including schools and clinics, and the appropriate administrative structure to coordinate all of this activity.

A successful plan requires the skills of many professionals, including town planners, architects, and social scientists who are familiar with local conditions; it also requires some degree of consultation with the people concerned, providing that this is done on the basis of definite proposals, and does not consist of conversations during which it is hoped to arrive at a solution. To indicate the complexity of the problem, let us consider some of the interrelated questions that must be asked before building new sites.

*Size and location of new settlements*

Agriculture is the basis of support of most settlers involved in relocation in connection with African man-made lakes. It is also intricately interrelated with the whole process of development. For this reason the location and size of the new settlements should be carefully thought out in terms of the productivity of available and accessible land resources under different systems of land use.

To date most resettlement has involved some increase in population densities. Where this occurred at Kariba it was primarily because there was insufficient land to reproduce existing spatial arrangements. Since no major attempt was made to intensify agricultural production, and since most resettlement soils were less fertile than the former alluvial gardens of the Tonga, some relocated communities have already used up the available land. Without major changes in land use or in employment opportunities, their plight in the years ahead will be serious. In Egypt, the Sudan and Ghana, the evacuees were concentrated because developmental opportunities were seen as greater with a higher density. In Ghana, the total number of settlements was reduced from 740 to 52, with some authorities—particularly the public health officer—pleading for an even greater concentration in the interests of development. Though resettlement here was at least accompanied by an ambitious attempt to intensify agriculture, it remains to be seen whether or not the new farming systems, once under way, will in fact be able to support the people at what they consider to be an adequate standard of living. At present the evidence is far from encouraging. In Egyptian Nubia the people formerly lived in communities which were spread along the Nile for approximately 300 kilometres between Aswan and the Sudanese border. Following relocation they were resettled north of Aswan, 'in planned, compact, contiguous settlements' which were concentrated in an area about 50 kilometres in length[24]. The situation in the Sudan was generally similar. In both cases, however, settlement size appears to have been carefully correlated with the capacities of irrigation systems planned in connection with resettlement. Though the new land unfortunately was not ready at the time of relocation, in the long run there is every reason to suppose that the people will become part of a more productive economy.

In deciding the size and location of new settlements, planners face the dilemma of acceding entirely to the people's wishes, and perhaps foregoing optimum possibilities for development, or of trying to impose what they consider to be a rational plan at the risk of antagonising the people. In Egypt and the Sudan such planning was entirely in the hands of government. In Ghana there was a lengthy process of negotiation and consultation between government and people over the choice of farm lands. The VRA wished to settle people on sites that had been selected for their availability, accessibility, and for their physical endowments—especially the presence of water and good soils. The people were also influenced by social factors, such as a reluctance to join a new settlement with a group to which they were traditionally opposed, or a desire to return to land to which they had some shadowy ancestral claim. The final 52 sites were chosen as a result of some compromises:

Twenty-five sites were selected by the people and accepted by the VRA;

Six sites were selected by the VRA and accepted by the people;

Sixteen sites were selected by the people and forced on the VRA;

Five sites were selected by the VRA and forced on the people[25].

While we have argued elsewhere in this chapter for an increase in the degree of consultation with the people, in some situations compromise may be impossible and consultation unwise. This may have been the case in the Egyptian and Sudanese cases noted above, where a number of reasons made it impossible to resettle the people along the lake shore margin.

*Public Works and Services in the New Settlements*

Throughout this chapter we have emphasised that resettlement, to be effective, must be viewed as a long term process with the responsibilities of the government and the settlers worked out well in advance. This is especially important where public works and services are involved. Even where relocation is not seen as a major opportunity to plan more productive environments (through the provision of more and better schools, health facilities, local government and security, market and trade facilities, etc.), it usually requires the construction of many miles of new access roads and the provision of water supplies. This is simply because empty areas are usually devoid of settlement for a reason, such as isolation, lack of water, and presence of tsetse flies or other pests which threaten the well-being of people and their stock. At Kariba the Southern Rhodesian government constructed 962 miles of access roads to the relocation areas. They also built 18 dams and two causeway weirs, and sank over 200 boreholes, the successful ones being fitted with pumps. The Northern Rhodesian government was involved in a similar effort, while both governments waged a relatively successful programme against tsetse fly by heavy ground or aerial spraying. These new facilities, of course, need to be maintained and serviced. Pipes and pumps must eventually be replaced. Tsetse fly must be actively contained to prevent re-infestation. It is unrealistic to expect the local offices of the relevant government ministries and the local authorities of the evacuees to take over immediate responsibility for such activities. Rather there must be a transition period during which the relocation authority gradually hands over its responsibilities. The length of this period will vary according to the circumstances: it is, however, extremely doubtful if it can be reduced below five years.

To date all the major resettlement schemes have been plagued with interruptions in major water supplies during the period following resettlement, while in Ghana, where inadequate provision was made for the upkeep of access roads, many fell into disuse with the onset of the first rains.

*Housing*

At Kariba the people were expected to rebuild their own villages, government assistance being restricted to transporting customary building materials (thatch and poles, in particular) which were scarce in certain relocation areas. This procedure has not been followed at Volta, Aswan and more recently at Kainji, where the tendency has been to construct housing for settlers. The provision of houses is undoubtedly part of a general trend toward a greater degree of state planning, an indication that the respective governments consider their duties to their citizens not fulfilled unless they build 'improved' housing for them—despite this being always one of the largest items of expenditure that is incurred in connection with resettlement.

Where housing is provided it is important that it be acceptable not only to the government but also to the settlers. Here there is bound to be some compromise solution between the people's wishes and the authorities who may be torn between a desire to reduce expenditures on the one hand and to do the right thing by the settlers on the other. In order to arrive at such a compromise the town planners should not only be familiar with the different types of housing prior to relocation, but also have an understanding of the practical, aesthetic and social reasons why the housing is built the way it is. Such information can be provided in at least two complementary ways. One is through surveys; the other is through the construction of experimental houses on which the settlers' comments are carefully evaluated.

The response from settlers to housing provided by the government has not been as favourable as it could have been with better planning. While this does not necessarily mean that the settlers are worse off (in fact, there is every indication that those relocated at Kom Ombo and Khasm el Girba will be considerably better off in the not too distant future), it does mean that they could have been provided with more congenial housing without sacrificing government intentions. We have already noted the Nubian objections to the spatial arrangements of their Kom Ombo housing. According to Fernea and Kennedy[26], 'There is scarcely a neighbourhood in New Nubia in which some houses have not been radically altered through the mounting of china plates above the doors, as in Old Nubia, and by plastering the exterior with mud to create a façade upon which traditional Nubian designs may be painted. *Mastabas,* the low clay benches running along the front of all Old Nubian houses, have also been added by many people . . . Less noticeable but more costly are interior structural alterations designed to separate human from animal quarters, to increase the enclosed area, and to organize the living area for greater efficiency and pleasure. Particularly important to the Nubians is the enlargement of space for entertaining visitors. Such house remodelling is costly. Some house-owners have spent as much as £E300 in their efforts to bring the new homes into conformity with traditional Nubian standards.' The Volta settlers also found their houses uncongenial and far too small for their needs, for here government built core houses consisting of one room each. Although there is provision for the settlers to add to the houses and make extra rooms, only a handful have done so to date, partly because the VRA insists on certain minimum standards that most settlers found too costly to implement.

At Kainji settler complaints can also be expected when they move into that portion of the new housing which was completed during 1966. Here the layout and design had been done by a local firm of architects which has had considerable experience in Northern Nigeria. While the architect responsible for this contract has produced some highly imaginative house types, his work has suffered because the necessary surveys had not been made in time for the results to be built into the town planning of the first major settlement. To be useful in the future, such surveys must be made well in advance. Since the rise in water level, after the dam is sealed, is sufficient to flood most riverine communities within the lake basin during the first few weeks, it is essential that the construction of new communities be completed at that time. This in itself is no easy task, especially where government is attempting to transform the people's environment.

*Local Councils and Community Development*

Among the many lessons of the TVA project was that the full development of all resources can be achieved only where there is effective participation from the people *and their institutions.* We stress this latter phrase to indicate the importance of having an administrative structure within which to promote development. But before considering the types of local councils and other administrative institutions, let us examine the prospects in a resettled area for community development which is sometimes proposed as an alternative to a formal local authority.

Community development (C.D.) from its formal inception in the mid-1950s has grown to become an almost universal phenomenon, with most developing countries having some sort of formal agency in charge of C.D. Rhodesia, Zambia, Egypt, Sudan, Ghana and Nigeria have trained C.D. workers, some of whom have been involved at one stage or another in the resettlement of the four lakes. They have encountered some formidable difficulties, as described by Peter du Sautoy, former director of C.D. and Social Welfare in Ghana. 'Since community development is concerned with the type of project that the people themselves really feel to be a need, and since to move away from their heritage of ancestral land is rarely one of the most pressing needs felt by human beings, community development practitioners are likely to face certain problems when asked to assist with resettlement schemes'[27]. C.D. is likely, however, to meet with *some* response, providing:

(a) there is no bitter legacy of wrangles with government over compensation;
(b) the idea of self-help has not been entirely replaced by dependence on government, such as led villagers in Ghana to refer ironically to the VRA as the 'Free R.A.';
(c) C.D. staff includes local people;
(d) the people have been involved in decisions from an early stage;
(e) some of the people can be persuaded to see the opportunities for improvement in their standard of living and to work toward achieving these opportunities;
(f) C.D. officials take maximum advantage of the receptive mood that may occur immediately after relocation, when initial hostility and suspicion are replaced by a passivity that can be turned to positive account.

Even if all these conditions are met, C.D. will be effective and lasting only if there is an organized administrative structure eventually to replace C.D.; for self-help and voluntary labour are not sound bases for economic development, whatever their romantic appeal.

The six countries which have been responsible for resettlement from the four major lakes all have more or less viable systems of local government. In Northern Nigeria, the emirates are the effective units, though district and village councils are emerging; Ghana is trying to reorganize her local authorities after years of abuses by the C.P.P.; Sudan provides for the granting of full local powers only when the area is ripe for local government, and in both Sudan and Egypt the central government exercises rather rigid control. In Zambia strong local authorities are encouraged, while in Rhodesia the exclusion of Africans from participation in national political affairs makes it difficult to engage Africans in local affairs. Despite differences of emphasis in the six countries, they have much in common in their approach to local government, partly because four (Zambia, Sudan, Ghana, Nigeria) were formerly administered by Great Britain and share the same local government traditions.

At Kariba, the Southern Rhodesian government has not attempted to develop local councils among the Tonga; rather, the District Commissioner deals directly with government supported chiefs who may or may not represent the interests of the population. By contrast, resettlement greatly strengthened the Local Authority of the Northern Rhodesian (now Zambian) Tonga, which was first established through government initiative in the 1930s. From the start the Administration required the Local Authority to support the resettlement programme, with the chiefs being 'compelled to use their powers to punish those who refused to submit'[28]. Following relocation, the Local Authority received £500,000 in compensation, which transformed it from being one of the poorest in the country to being the richest. Additional revenue came from the commercial fisheries that developed on Lake Kariba, and from increased trade in beer. By 1963, area councils had been established within the Local Authority in order to encourage local participation.

While the major problem facing the creation of a strong Tonga Local Authority has been the general absence of any form of strong political organization extending beyond the boundaries of a cluster of villages, in Ghana the ethnic diversity of the evacuees and the policy of merging new settlements with old ones has presented difficulties. When merged with existing communities, many bitterly resisted the loss of their own identity as expressed in traditional councils or local authorities. This applied strongly to some of the smaller groups whose insecurity and anxiety at loss of identity were the greatest. Though the population at Kainji is also ethnically complex, the situation here is quite different because of the existence of strong emirates which have extended their authority over all the communities within their boundaries.

There are, of course, many difficulties in attempting to set up strong local councils, only some of which we have mentioned. Though a revenue base is essential, settlers may complain (and with some justification at first) that they cannot afford to pay taxes. Then there are the standard problems of local government; of numbers of people; size of area; recruiting, training and keeping responsible, efficient and honest officials; involving the people at the community level; supervision and guidance, but not stifling control, from central government; development of a capital account; deciding the proportion of expenditure for development, and many others. Because ultimately a stable and efficient form of local government is one of the most effective preconditions of development, it should be encouraged in every way.

Other formal administrative structures, such as cooperative societies to help organize agricultural production, fishing, or marketing, may also be appropriate, provided their conception is based on careful planning and on an adequate understanding of the local situation.

*Social Developments*

Once some satisfactory form of administrative structure is created, it will be easier to establish and maintain facilities for improving health and education, both of which are integral parts of any comprehensive plan for resource development. We stress the prior need for administration, because coordination is essential in these fields, and much energy and money can be dissipated in piece-meal attempts to build another class-room, encourage the people to use latrines, or set up a dispensary. For lasting effects, a framework such as that provided by a good local council, with its links to the specialised agencies of central government, is essential.

EVACUATION

The success of this part of resettlement depends in large measure on the degree and quality of the contact previously made with the people by the government, and on the coordination of the physical move. The physical problems of evacuation are tremendous, because the people, livestock, household effects and food have to be moved from perhaps hundreds of scattered, isolated and inaccessible sites within a short time. At Volta 'the settlers were accompanied by 2,954 cattle, 11,600 chickens and 42,000 sheep and goats'[29]. The Sudanese Nubians were accompanied by an average of one thousand pounds of baggage per person. Despite the volume of possessions, it is not desirable to attempt to dissuade people from taking all their belongings, even if many appear to be worn-out by age and use, as this would only increase the trauma of leaving their homes, most of all for old people. Given the bulk of people, livestock and possessions, elaborate transport arrangements must be made. This means not only the provision for boats and the right sort of heavy-duty vehicles, but also ensuring that they are in good working order

and that there are facilities for their maintenance. In several instances evacuation has been delayed for lack of mechanics to repair broken-down trucks. Roads have to be built, for some of the villages are bound to be inaccessible. It may be necessary to obtain fleets of boats, as at Aswan, or to use amphibious vehicles and motor launches and canoes, as at Volta. Once again, detailed prior organization and coordination are essential if the evacuation is to be handled smoothly.

At Volta the national army helped to move thousands of people. It was considered an obvious source of help since it had access to a large fleet of well-maintained vehicles, and since army officers are accustomed to perfecting the elaborate logistics involved in such exercises, and to putting them into effect with dispatch. Use of the army, however, can introduce further complications if there is a popular anti-military feeling, or if the army proves impatient with waverers, or irritated by those who are not ready to move, and wishes simply to use force as a solution. The resettlement officials, as happened in Ghana, then have to persuade the army that a little patience and latitude at this stage will ensure goodwill later, and that resettlement is an ongoing exercise, not a crash programme.

The VRA team of resettlement officers went to great lengths to avoid force, threats or punitive action, perhaps because Southern Ghana has a markedly hortatory rather than a violent tradition. 'If you disagree with someone, you talk to him and try to get him to change his ways', say Ghanaians, and this is what happened during the evacuation from the Volta. Some old people refused to move until the rising water had actually reached their houses. Because force was ruled out, the VRA had to use more vehicles —and boats—than it would had the move been conducted as a military operation, but the humouring of the people in this case certainly led to the establishment of good relations. At one point, when the evacuation had reached a peak, the principal settlement officer received a telegram from an assistant saying '90 old lady unable walk please send taxi'. Puzzled but willing to help, the officers commandeered all the local taxis and sent them to the village to collect the ninety ladies, only to find that the telegram had been mutilated and should have read 'One 90 *year* old lady unable walk, please send taxi'. The outcome was that all the people of that settlement were able to ride grandly out in the fleet of taxis.

An ironic commentary on the Kariba evacuation was that the plight of the wild animals, which were marooned on the islands and rescued in 'Operation Noah', received much more attention than the fate of the 50,000 human beings. There was certainly much publicity in the form of articles, books and films, given to the baboons, antelope and larger game. On the Volta lake there were far fewer animals, which were anaesthetised and relocated quietly and with no attendant publicity.

In cases where the people have been moved short distances to their new homes well in advance of the time of inundation, there is a tendency for some to slip back to their old homes, for sentiment or to rescue a few belongings, for ceremonial purposes or because of a doubt that the water will rise. Whatever the reason, this is another potential problem for the resettlement authority.

### INSURANCE OF FOOD SUPPLY

None of the relocation areas associated with the Kariba, Volta and Aswan High Dam projects were capable of supporting the people at the time of relocation. At Kariba, the Southern Rhodesian Government provided grain, dried milk and iodized salt during the most difficult months of the first two years following relocation. In the Sudan and Ghana the governments requested and received the assistance of the World Food Programme. Regardless of the arrangements made, the provision of food relief should be the responsibility of the resettlement authority until the evacuees are once again self-sufficient. Responsibility here implies having the right kinds of foods in the proper quantity available at the right places and at the right times. In Ghana most people are still on food relief one or two years after relocation. Under such circumstances, an efficient storage and distribution system is essential if the people (and especially the very young and the very old) are not to suffer.

Extended periods of food relief are expensive both in terms of food imports and in terms of the loss of productivity of the evacuees. Food relief also has a disheartening and demoralising psychological effect on the settlers. In any resettlement programme the responsible authority has to do much for the people simply because there are many tasks which are beyond their ability at the time. Under such circumstances there is a danger of building up too great a dependency on the government and of discouraging local initiative. These dangers are aggravated by prolonged periods of food relief.

The best corrective is better planning, timing and coordination of relocation so that the new economic base prepared for or by the people is ready to receive them as producers at the time of resettlement. Whatever additional costs are involved in better planning will be more than offset by the short-term gains in the period immediately following resettlement.

## HEALTH

We have already dealt with the need for health surveys in an earlier section. Though the statistical evidence is not available, the stress that accompanies resettlement, the often inadequate feeding of the population prior to the functioning of new economic systems, and a scattering of medical reports all strongly suggest that mortality and morbidity rates do increase during the months following resettlement unless special precautions are taken. Among the Nubians relocated at Kom Ombo, 'communicable diseases such as dysentery, measles and a form of encephalitis quickly spread in the suddenly condensed population. These conditions, aggravated by the high summer temperatures typical of the region, caused a rapid rise in mortality, specially among the very young and the very old'[30]. At Kariba, sleeping sickness (trypanosomiasis), apparently in epidemic form, was serious in at least one Southern Rhodesian resettlement area. North of the Zambesi, 'of 1,600 Tonga who were moved from the Valley Floor to the Siagatuba area on the adjacent Plateau during 1958, forty-one children died during the first three months of 1959 . . . . As for the 6,000 Tonga who were moved some 100 miles down-river to the Lusitu basin, up to 100 people (mostly children) died during the months immediately following resettlement, with a majority of the deaths attributed to bacillary dysentery. A year later fifty-three women and children died of "an acute condition of sudden onset and high and rapid mortality"[31]. These deaths were concentrated in about eight villages, one of which lost approximately 10 per cent of its population'[32].

In most resettlement schemes, the health of the people has improved after the first or second years. Our concern here is with the stress period immediately following resettlement. In some cases lasting as long as several years, this is the time when the people are most susceptible to illness. Yet this is also the time when they are expected to settle into (and sometimes build) their new homes and to introduce certain planned innovations into their lives. Obviously it is important to have as healthy a population as possible during this important phase, hence the need for comprehensive health surveys prior to relocation and better surveillance during and immediately following resettlement.

RESEARCH

Man-made lakes offer an exceptional opportunity to undertake fundamental research which can be expected to increase man's ability to control and intelligently manipulate his total environment. 'Rates of change are accelerated through such massive human interference, and studies that are carried out before, during and after inundation can be expected not only to increase knowledge, but also to contribute directly to planned development. In other words, the dichotomy between pure and applied research simply does not apply to such man-made situations. This is true whether the investigator is a geophysicist studying possible crustal movement arising from the added weight of millions of acre-feet of water, or a behavioural scientist interested in the effects of relocation on human populations. We really know very little about how people will behave under different types of resettlement, but population relocation presents us with an important opportunity to widen our knowledge of human behaviour under conditions of increased stress. Such knowledge will also increase our ability to induce the type of social change that is meaningful in terms of both individuals and the societies to which they belong'[33].

Research in connection with African man-made lakes has been totally inadequate to date. This is especially true in regard to the people undergoing relocation. With millions of people being relocated throughout the world every year, there is a tremendous need for what one of us has called 'a sociology of resettlement'[34]. What is needed is a series of long-term systematic enquiries that have as their main aim an increased understanding of the processes of social change. Also needed are broad-based economic studies of the major dam projects which evaluate their costs and benefits at different points in time, and compare these with alternative plans that might have been initiated either by funding an entirely different project or series of projects or by altering certain aspects of the project concerned. Such studies should, of course, include the costs and benefits of population relocation and the development of the lake basin. While there are many intangible factors involved which the economist should take into account, the task is by no means impossible, as shown by recent cost-benefit investigations of education. While such research in the long run will enable subsequent resettlement projects to be better planned and implemented, it must be distinguished from social and other surveys which are planned for specific projects and designed to provide the type of information planners need when they need it.

To date the initiation, staffing and financing of such long-term studies has come on the one hand from local universities and research institutions, and on the other from international and other external agencies. Recently the Nigerian Institute of Social and Economic Research has played an important role in initiating research in connection with the Kainji Dam scheme. This is designed to provide survey information for those responsible for planning and to provide the baseline information for long-term studies. At Kariba the Rhodes-Livingstone Institute for Social Research (now the Institute for Social Research of the University of Zambia) was responsible for two detailed studies of the Gwembe Tonga before and after their relocation. As for external agencies, the Social Research Centre of the American University in Cairo carried out detailed baseline studies prior to the resettlement of Egypt's Nubian population in connection with the Aswan High Dam project. More recently, the UNDP-FAO-host country series of African man-made lakes projects has been broadened to include the sociology and economics of resettlement. The trend here is encouraging, though it remains to be seen whether or not the studies initiated receive sufficient support in the years ahead.

Another encouraging trend is the increasing number of individual scholars, universities and other institutions, both within and outside the countries concerned, that are becoming interested in research in connection with African man-made lakes. To deal with this situation, it is vital, we believe, that each host country establish under local leadership an administrative structure which on the one hand can set research priorities and on the other coordinate existing research with the activities and interests of the river basin authority and other interested government agencies. The University of Ghana formed a Volta Basin Research Project which aimed to coordinate all research, to prevent overlap and to provide assistance and support to those interested in making enquiries. For financial and other reasons very little research was accomplished in the social sciences. At Kainji research has been organized along more rational and effective lines, though here the organization took place after the initiation of dam construction, and much of the desired research has yet to start. The organization is in the form of a National Advisory Committee, under the chairmanship of the Minister for Agriculture and National Resources. Members are to be drawn from National and Northern Nigerian government agencies with a Kainji interest, the universities and research institutes, the Niger Dams Authority, and the United Nations Development Programme.

An important benefit of both social surveys and long-term research is the opportunity for providing field training to students. Part of the training of any social scientist consists of exercises in formulating and administering questionnaires, conducting social surveys and other enquiries, and being exposed to field conditions. Students in developing countries should have more experience in supervised field research, for several reasons. First, it serves as an apprenticeship, to prepare them to undertake later independent studies on their own. Second, it should help them relate their academic studies to the world around them, for there are real dangers of any university becoming an ivory tower or being thought to be so. Third, there is a strong tendency for many students to become increasingly removed from the people of their own country, and to regard themselves as an intellectual élite, having little in common with those who have had less education. In these circumstances, the students are not likely to have any interest in, or sympathy with, the problems of the rural people. If this attitude prevails, it will be a hindrance to development, in that the future academic leaders will not appreciate the importance and the potential, in development terms, of the people who make up more than 85 per cent of the population. Nor will they appreciate the opportunties for research.

To aid those involved in long-term research as well as surveys in connection with man-made lakes, and more specifically with population relocation, an international clearing house for the collection, classification and distribution of relevant information is needed. At present most reports dealing with lake basin resettlement and development are not published, so that no matter how relevant they are they are rarely available to personnel concerned with similar projects in other countries.

Though outside our topic, we wish nonetheless to stress in this chapter the need for archaeological research in connection with African man-made lakes. Though excellent work has been done along the Nile in connection with the Aswan High Dam Project, and some was done before the filling of Lake Kariba, archaeological investigations were not undertaken along the Volta and have yet to occur along that portion of the Niger to be flooded by the Kainji project. This is unfortunate, especially since river valleys are important sources of African history. Before inundation, the various sites should be mapped and the more important ones excavated. This is especially relevant in Africa where governments are playing an active role in attempting to discover more about the cultural history of their peoples. Furthermore, a better knowledge of the past may even aid planning for the future by broadening our understanding of man's relationship to his environment.

## SUMMARY

Throughout this chapter we have emphasised that resettlement should be regarded as an integrated part of a multi-purpose river basin development project. This means that all aspects of resettlement should be directed with development in mind. They should be seen as integral parts of a long term project, and not as piecemeal emergency measures.

We have also stressed that there is no single way to handle resettlement because of the wide variation in political ideology and administrative policy, as well as in natural and human resources. People are a country's major resource; they can be remarkably resilient and show a great capacity of accommodation to new conditions. In connection with resettlement, this is illustrated by the rapid build-up of the Kariba Lake fisheries a year after relocation, and the appearance of new customs and attitudes, especially a new optimism, among the Nubians resettled at Kom Ombo in Egypt.

Resettlement *is* a formidable task, even under the most favourable conditions. Just as governments have tended to underestimate the potential of populations undergoing relocation, so have they underestimated the difficulties involved in preparing new areas for large numbers of people (often divided into ethnically distinct populations) and in physically relocating them according to a rigid time table. Since the original decision to build a major dam is the government's decision, government has no choice but to take the initiative in trying to achieve successful resettlement. The opportunities offered for development are great; but they can be achieved only through careful planning, timing and utilisation of a wide range of available sources of assistance.

## BIBLIOGRAPHY


A.I.D. (Agency for International Development), *Evaluations, Suggestions, and Recommendations on Resettlement of Volta River Area, Ghana,* Washington D.C., 1963.

David Brokensha, 'Volta Resettlement and Anthropological Research', *Human Organization,* 22, 1963, pp. 286–290.

Elizabeth Colson, *The Social Organization of the Gwembe Tonga,* (especially Appendices C and D on resettlement in Northern and Southern Rhodesia respectively), Manchester, 1960.

........................, 'Social Change and the Gwembe Tonga', *Rhodes-Livingstone Journal,* XXXV, 1964, pp. 1–13.

Peter du Sautoy, 'Resettlement Schemes and Community Development', *Community Development Bulletin,* XII, 1961, pp. 121–123.

Robert A. Fernea and John G. Kennedy, 'Initial Adaptations to Resettlement: A new Life for Egyptian Nubians', *Current Anthropology,* VII, 1966, pp. 349–354.

K. G. Gadd, L. C. Nixon, E. Taube and M. H. Webster, 'The Lusitu Tragedy, *Central African Journal of Medicine,* Supplement No. 8, 1962.

Alan W. Horton, 'The Egyptian Nubians', *The American Universities Field Staff Reports Service,* Northwest Series, XI, No. 2 (Egypt), 1964, pp. 283–302.

G. W. Lawson, 'Volta Basin Research Project', *Nature,* 199, No. 4896, 1963, pp. 858–859.

Rowena Lawson, 'An interim appraisal of the Volta Resettlement', unpublished paper, 1966.

David E. Lilienthal, *TVA: Democracy on the March,* New York, 1944.

Rosemary McConnell and E. B. Worthington, 'Manmade Lakes', *Nature,* 208, No. 5051, 1965, pp. 1039–1042.

Thayer Scudder, 'The Kariba Case: Manmade Lakes and Resource Development in Africa', *Bulletin of Atomic Scientists,* 1965, pp. 6–11.

........................, 'Manmade Lakes and Social Change', *Engineering and Science,* XXIX, 1966a, pp. 19–22.

........................, 'Manmade Lakes and Population Resettlement in Africa', in *Manmade Lakes,* ed. Rosemary McConnell, London, 1966b.

V.R.A. (Volta River Authority), *Volta Resettlement Symposium Papers,* Kumasi, 1965.

[1] Scudder, 1965, p.7.

[2] Scudder, 1966a, p.22.

[3] Scudder, 1966a, p.19.

[4] Lilienthal, 1944, pp. 62, 64.

[5] *Ibid,* p.125.

[8] See Scudder, 1966b, pp. 103–4

[7] Lawson, 1966, p.7.

[8] See Scudder, 1966b, pp. 103–4

[9] Horton, 1964, p.12.

[10] VRA, 1965, p.13.

[11] See AID, 1963, pp. 3, 27.

[12] Horton, 1964, p.12.

[13] VRA, 1965, pp. 66–72.

[14] VRA, 1965, pp. 147–156.

[15] Scudder, 1966b, p. 104.

[16] There are several field manuals that cover the framework that should be used for this sort of enquiry. See for example, M. G. Smith and G. J. Kruijer, *A Sociological Manual for Extension Workers in the Caribbean,* Univ. College of West Indies, 1957; also Hsin-Pao Yang, *Fact Finding with Rural People: A Guide to Effective Social Survey,* Rome, F.A.O., 1955.

[17] VRA, 1966, p.19.

[18] Fernea and Kennedy, 1966, p.352.

[19] Scudder, 1966a, p.20.

[20] VRA, 1965, p.22.

[21] 1966, p.350.

[22] Horton, 1964, p.14.

[23] VRA, 1963, p.53.

[24] Fernea and Kennedy, 1966, p.349.

[25] VRA, 1965, pp. 19–51.

[26] 1966, p.350.

[27] du Sautoy, 1961, p.121.

[28] Colson, 1964, p.9.

[29] VRA, 1965, p.19.

[30] Fernea and Kennedy, 1966, p.350.

[31] Gadd et al, 1962.

[32] Scudder, 1966b, p.104.

[33] Scudder, 1966a, p.22.

[34] Brokensha, 1963, p.286.

CHAPTER IV

# The Volta Dam: Its Prospects and Problems

ROBERT W. STEEL[1]

THE Volta River Project may well be the greatest single project that has so far been undertaken in tropical Africa, and it is certainly the biggest single step ever taken in the economic and industrial development of Ghana. It is a scheme with many sides to it, and in the view of Professor Arthur Lewis—a person with tremendous experience of the problems of developing countries—it will bring to Ghana a larger income per head of population than any other industry or enterprise in which Ghanaian labour and money could be employed. It may also put Ghana in the forefront of the industrialised parts of tropical Africa, and thereby minimise the country's present state of being dependent upon the production of a single crop, cocoa.

## THE HISTORY OF THE PROJECT

The Volta Project was the central point—or perhaps it is better to say the cornerstone—of the Seven-Year Development Plan, designed to cover the period from 1963 to 1970. Thus it became very much the focus of a great deal of national aspiration and sentiment during the decade after the Gold Coast became independent as the State of Ghana. But it is equally important, that we should recognise that it is a project with a considerable history, most of it pre-independence. The Gold Coast Government was considering its development in now far-off colonial days, as long ago as in 1924. Even then the possibilities of exploiting the country's extensive supplies of bauxite—first discovered in 1915 by Sir Arthur Kitson, Director of the Geological Survey of the Gold Coast—were being considered. During the latter part of the inter-war period possible plans for development were drawn up by Mr. Duncan Rose, a South African engineer who conceived an integrated hydro-electric and aluminium project that bears very close resemblance to what in fact has now been established. At the end of the Second World War, there was considerable unofficial interest in the possibilities of a hydro-electric development on the lower Volta, and a company called West Africa Aluminium Ltd. was formed to carry forward the scheme. The government of the Gold Coast first took an active part in 1949 when the London firm of Sir William Halcrow and Partners was commissioned to investigate the prospects of a Volta scheme in relation to the development of the country's general economy. The Halcrow report, presented in 1951, was favourable and made clear the scale of the financial commitment that might be involved: and the upshot was a White Paper (Cmd. 8702, published in November 1952) and the establishment during the following year of a Preparatory Commission under the chairmanship of Commander (later Sir Robert) Jackson.

In the years that followed there were many complications. Alongside growing governmental involvement was an understandable hesitation on the part of certain Gold Coast leaders who feared that this might be the prelude to a new onslaught and invasion from western capitalism. An independent government-commissioned report was prepared by Professor Arthur Lewis who was then at the University of Manchester, and the Preparatory Commission produced its three large and valuable reports in 1956. Ghana came into being a few months later in March 1957 and the new government immediately sought to interest others in the financing of the project. Unfortunately there was almost simultaneously a marked fall in the price of, and in the demand for, aluminium; and in some quarters doubts were expressed about the political and financial stability of the new government. In order to keep the issue alive, and to make it economically more viable and financially a little less daunting, a new technical assessment was sought, on behalf of the government of Ghana, from the Kaiser Corporation of California.

The Kaiser Report modified the project in various ways and suggested some very important savings, including a new and more economical dam at Akosombo, two miles downstream from the dam-site originally selected at Ajena. The revised scheme was eventually accepted by all the different interests concerned—including the governments of the U.S.A. and of Britain in addition to that of Ghana, the International Bank for Reconstruction and Development, and the aluminium concerns who combined to form the Volta Aluminium Company Ltd. or VALCO. But before any outside finance was committed, the Ghana government agreed to finance all engineering and preliminary construction work for the four-year period 1958-62. The contract for the building of the dam itself was put out for international tender. In this way the government was involved in an expenditure of £G 25 million prior to the go-ahead given by President Kennedy in January 1961. The contract for the dam was signed in May 1961, and work was started two months later by the Italian consortium, Impregilo, who had already had the very valuable and relevant experience of building the Kariba dam on the Zambesi.

The idea is, therefore, quite an old one, but the implementation of it has been very recent and indeed is essentially one of the achievements of the independent government of the post-1957 period. It would not be unfair to regard it as one of President Nkrumah's most cherished dreams—certainly nothing less than his drive and enthusiasm could have carried the project through despite all the obstacles and difficulties of recent years. It is ironical indeed that the revolution that overthrew him and his government took place within a month of the inauguration of the Project on 23 January 1966.

This paper is not concerned with a technical assessment of the Project or with its engineering particulars, but it is necessary to indicate briefly the main features of the plan with a few details about the dam itself and the newly-created lake. The basic idea throughout has been to dam the Volta, one of West Africa's major rivers, somewhere in its lower reaches in order to generate hydro-electricity on a large scale. The purpose of this generation has always been directed towards the production of aluminium—a commodity that is generally increasingly in demand and one that requires considerable supplies of low-cost electricity. Each pound of aluminium requires 7 kilowatt hours of electricity, and most of the world's production facilities are sited where power costs are less than 0·4 pence per kilowatt hour. The metal is manufactured from bauxite, a raw material with which Ghana is generously endowed. Ghana could in fact become one of the world's major producers of bauxite: her present exports of 300,000 tons could easily be increased to 1·2m. tons per annum—an amount exceeded only by the U.S.S.R., Jamaica, Surinam (Dutch Guiana), Guyana (formerly British Guiana), the U.S.A. and France. Ghana was to become a leading producer of aluminium with the first-stage aluminium reduction plant at Tema beginning production during 1967, while the second stage of expansion will lead to

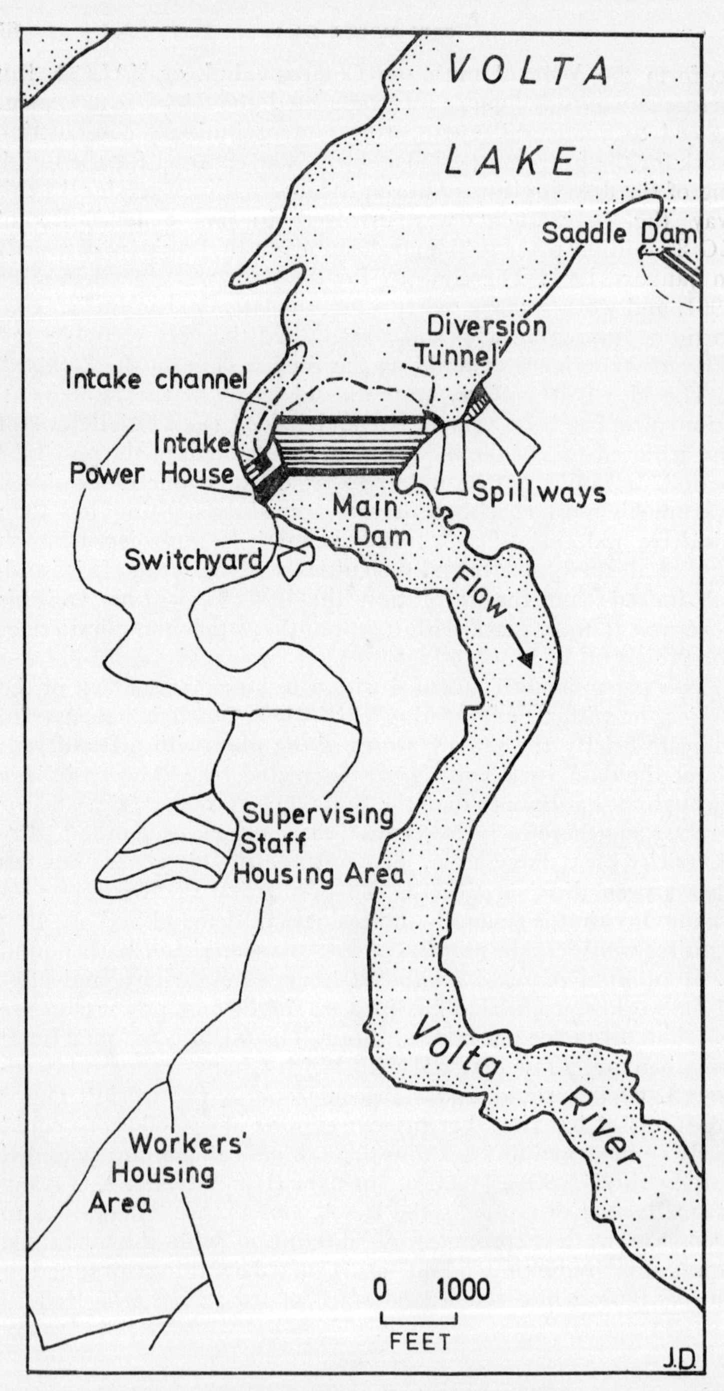


1. The Volta River Dam at Akosombo

an output of 200,000 tons by 1970. This will then be the largest-sized plant outside the U.S.S.R. and the U.S.A.

THE AKOSOMBO DAM

The dam has been built at Akosombo at a point where the river has a steep gradient as it cuts through the quartzite and sandstone ridges of the Akwapim hills to the plains of the lower Volta. It is designed to create a huge storage reservoir in order to stabilise over many years the fluctuations of the river which varies between a minimum of less than 1,000 cusecs (cubic feet per second) and a maximum of over 500,000 cusecs. The Volta Lake, when fully filled, will have a surface area of 3,275 square miles and a coastline 4,500 miles long and will stretch for 250 miles above the dam: it will be one of the largest man-created lakes in the world.

The dam is of the rockfill type and is 244 feet above river level and 370 feet above the foundations. It has a crest line of 2,100 feet. A subsidiary saddle dam is 120 feet high and nearly 1,200 feet long. Water began accumulating behind the dams prior to the flood season in the autumn of 1964. The power station was nearly completed during 1965, the first unit being commissioned on 17 September 1965: and Akosombo has supplied nearly all Ghana's electric transmission grid requirements since that date. The reservoir level filled to the normal minimum operating elevation of 248 feet during 1966, and the smelter at Tema was able to begin production early in 1967. The power station now has an initial capacity of 589 megawatts with four generator units and, with the installation of two further units, an ultimate capacity of 883 MW. Even the 1967 capacity is five times that of the installed capacity of all the electric plants in Ghana in 1961.

The Volta River Authority (VRA) is responsible for the construction of the dam, the provision of transmission lines, the building and maintenance of the hydro-electric station, and various other things like the compensation of those whose homes and land have been flooded by the new lake: the government of Ghana, on the other hand, is responsible for the resettlement and rehabilitation of the 70,000 people concerned. The cost of the VRA items is roughly £70 million: half has come from Ghana itself, about £18 million from three U.S.A. agencies, and £17 million as a loan from the World Bank. VALCO (the Volta Aluminium Company Ltd.), formed in 1959, a consortium of two American companies (90 per cent Kaiser Aluminium and 10 per cent Reynolds Metals), is responsible for the aluminium smelter, the cost of which is estimated to be £60 million. There are also other aspects of the project that underline the magnitude and the many-sided nature of the whole operation. Railways and roads have had to be built to make the construction of the dam possible, and still more will be needed if Ghana's supplies of bauxite are to be effectively used. Improved port facilities were vital for the Accra area and for the Eastern Region of Ghana in general, quite apart from the special demands and the new potentialities of the Volta Project itself. Hence the decision taken before independence to proceed with the new port of Tema, 18 miles east of Accra, whether or not the dam and smelter were built. The necessary rail and road construction was started in 1952 and work on the artificial port began in 1954. Tema was opened in 1962 and already has ten deep-water berths, with a total length of 6,000 feet. Its population, originally designed to be 80,000, is now planned to reach 250,000 within ten or fifteen years by which time Accra and Tema may well form a single conurbation with a total population of more than 1¼ millions. The town will have important industrial as well as commercial functions, and VALCO's smelter has been built there, 44 miles away from the dam: this represents a significant change from the earlier plans which recommended that the smelter should be at Kpong, very near to the dam itself.

### ASSETS AND PROSPECTS

It is now necessary to attempt some general assessment of the dam and of the Project as a whole, appreciating the tremendous potentialities while at the same time recognising the very considerable problems, technical, financial, social and industrial, that such developments present when they take place with almost dramatic suddenness within the area and the economy of a developing country.

First, what are the assets of the Project and what does it offer to Ghana? Of the advantages, the outstanding one is the provision of inexpensive electricity in large quantities, primarily for the aluminium industry, which will initially require one-third of Akosombo's generating capacity and, in the second stage, two-thirds. Aluminium manufacture involves two processes: first, there has to be the recovery from bauxite of commercially pure aluminium oxide or alumina by a process involving the use of caustic soda and energy; and secondly, alumina is reduced electrolytically to aluminium in specially designed cells. It is during the second stage that the energy requirement is particularly large

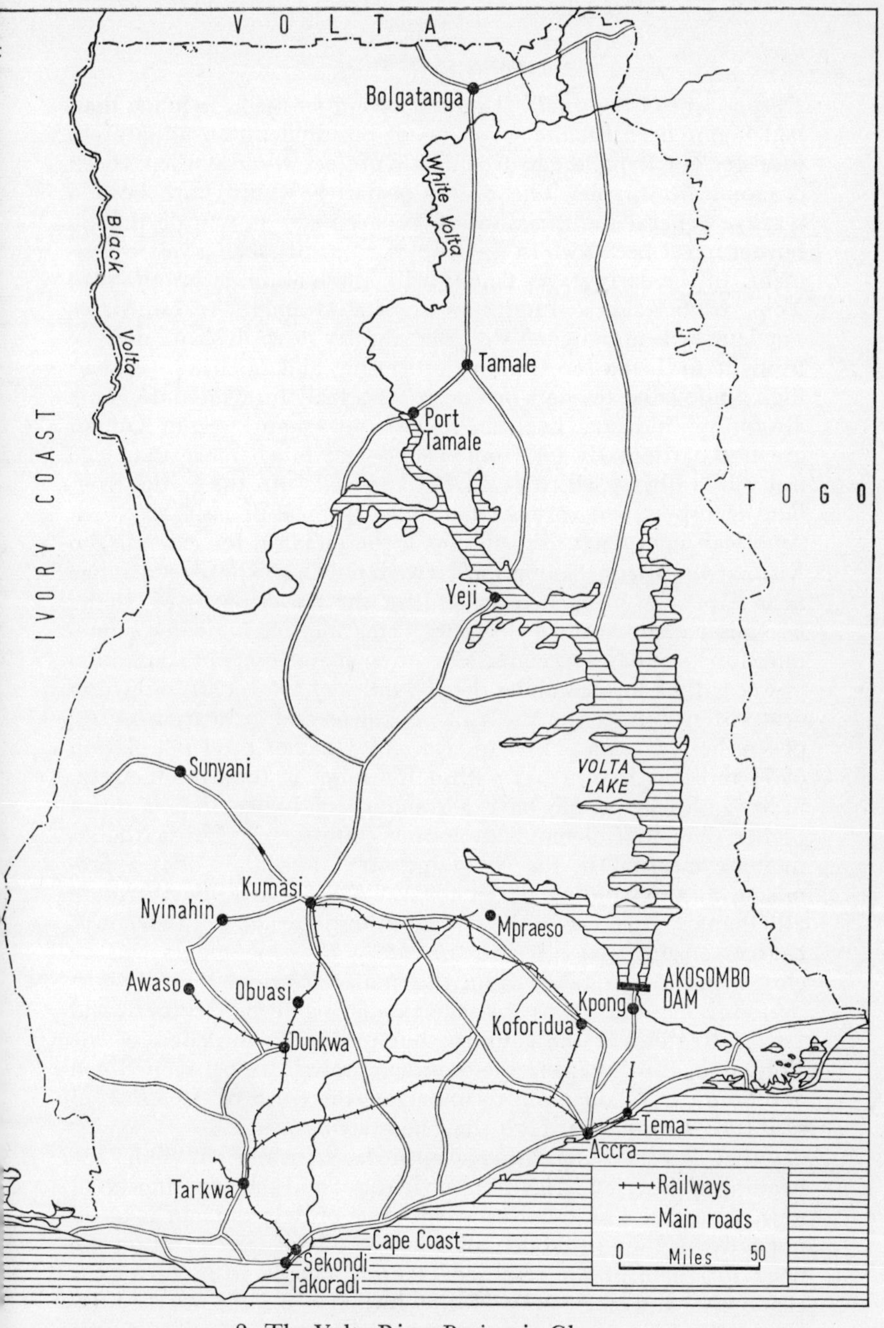


2. The Volta River Project in Ghana

*Bauxite is mined at Awaso which was opened up during the Second World War when a railway was constructed from the main line at Dunkwa. There are also bauxite deposits at Mpraeso and more extensive deposits to the west of Kumasi at Nyinahin.*

(90 per cent of the total). It cannot be emphasised too much that, had it not been for the prospects of building up an aluminium industry in Ghana, the hydro-electric project would not have been economically viable. The only alternative would have been a massive general industrialisation programme, on a scale that so far has never been seen in the tropics, to justify such a vast investment. In the early stages Ghana will refine alumina imported to Tema from Kaiser's aluminium plant at Gramercy in Louisiana, but later it is anticipated that the country's own bauxite may be brought to Tema for alumina processing and smelting. Ghana's high-grade bauxite offers prospects for a fully integrated domestic aluminium industry. Even the deposits in Ashanti west of Kumasi are not particularly far from Tema—not much more than 200 miles by railway, which is small compared with the initial overland transport and voyage from many tropical deposits to North American manufacturing sites, as from Jamaica, for example, to Kitimat in British Columbia. New deposits on the Afena range near Kibi now being investigated for the Government of Ghana by Kaiser engineers are closer to Tema and offer greater economies for local developments. The other major users of Akosombo power will be located along the 500 miles of 165 KV. transmission grid system connecting the growing towns and industrial centres of southern Ghana. Three-quarters of Ghana's total population of 7½ millions live in fact within 150 miles of the capital Accra. Many of these people have a standard of living that is much higher than that of most developing countries in Africa (thanks in large measure to the cocoa industry), and they offer a promising and expanding outlet for Volta electricity. Furthermore all Ghana's mines are in the same southern area, and lower power charges ensure that all their private generating plants will be closed and power taken from the national grid within a short period. Long-term plans envisage extensions to the Northern and Upper Regions of the country; but meantime the effect of the Volta dam is to intensify the economic imbalance between North and South in Ghana, and to underline the truth of the words of scripture 'to him that hath shall be given'.

Other aspects of the Project can be dealt with briefly under the headings: fisheries, irrigation, navigation, and the exploitation of new resources. The fisheries programme of development is based not only on the availability of a large body of water and on the need for some of the resettled lakeside people to modify their economies and their way of life but also on the marked inadequacy of the protein supply in many Ghanaian diets. Cattle cannot be reared in many of the tsetse-infested parts of the country: those that are kept produce meat that is sometimes prohibitively expensive for the ordinary person. Whereas up till now there has been only limited and local fishing on the Volta river, it is now planned to obtain 18,000 tons a year—the equivalent of the whole offshore fisheries catch at the present time. About 6,000 tons of fish per annum are already being landed, and it is estimated that 40,000 canoes are engaged in fishing.

Irrigation is often associated with major hydro-electric projects. The Accra Plains lend themselves to such developments since they are some of the driest areas of the whole of the Guinea coastlands. Increased and intensified cultivation is one possibility, and there are some encouraging experiments being carried out at Kpong with rice and sugar, both at present among the chief food imports of Ghana. Two sugar schemes are being developed with water pumped from the lower Volta. A study of the entire area of the Plains prepared by Kaiser Engineers suggests that there might be 800,000 acres irrigable. A contract has been signed by the Government for the carrying out of detailed surveys of the first tract of 22,000 acres and for the first stage of a high canal which will eventually extend from a pumping station at Kpong to Accra and will provide not only irrigation water but also urban water supplies for Accra and Tema. Water supplies in quantity for livestock may offer even better prospects, since the Accra Plains area is immediately adjacent to the largest, most affluent and most rapidly increasing market in the whole of Ghana. There are numerous problems of topography and of soil types, but the overall potentialities are good, and extensive uncultivated and uninhabited areas may eventually be settled and developed as a direct result of the new availability of controlled supplies of water from the Volta Lake.

Navigation on an artificial lake the size of the Volta Lake clearly offers possibilities for the development of water traffic, especially since the dominant trend of the lake is from south to north. There is, of course, no such traffic already in existence, and the immediate surroundings of the lake do not support a large population or produce appreciable quantities of foodstuffs, minerals and other commodities. Indeed the immediate effect of the creation of the lake is the disruption of some of the existing ferry services at places like Yeji, where the 'Great North Road' linking Kumasi with Tamale and the North has always crossed the river. Larger and better-equipped ferry services are being supplied as part of the project, and congestion and delay may be minimised as a result. The Development Plan goes much farther, however, and sees the lake as opening up new areas with reasonable resources. It also believes that the lake can become 'a major new artery of communication which it is proposed to incorporate in a rationalized transport system for the country as a whole' (*Plan*, p. 211). It envisages the building of lake ports and the establishment of a transport system that could eventually handle much of the existing north-south traffic. The system would involve a rail roll-on–roll-off barge system on Lake Volta linked by a 23-mile railway to the existing main-line system of southern Ghana at Koforidua from a new port at Apimso on the lake. Each barge would convey forty cars on four tracks over a distance of 220 miles and at a speed of 10 knots to Port Tamale. There a railway of 27 miles could be built to Tamale, whence further extensions might be made northward to Bolgatanga, Ouagadougou and Niamey. This would bring thousands of square miles within the economic trading orbit of Ghana, and the whole system would be much more economical than the existing main roads in this vast, and as yet scarcely developed, area in the interior of West Africa[2].

The exploitation of new resources is visualised by the planners with some confidence in the belief that northern Ghana may—on the analogy of Nigeria—have considerable possibilities if only the problems of transport and accessibility can be solved. Certainly agricultural and fisheries developments are possible, and there are limestone deposits at Buipe and iron ore at Shieni: the latter, it has been suggested, could become with Ghana's manganese the basis of a ferro-manganese industry using electro-manganese reduction.

However large or small such new developments may be, the effect on the economy as a whole is bound to be considerable in the aggregate. If all the items on the credit side are added together, beginning with electrical power and bauxite and ending with the more remote possibilities just referred to, the broadening of the basis of the economy of the whole country is obvious. Up to now agriculture has been far too dominant. Two-fifths of the population is thought to be directly concerned with the cultivation, harvesting and marketing of cocoa—a quarter of the people probably have cocoa-farming as their major occupation. Many more (up to two-thirds of the total) are chiefly concerned with the production of foodstuffs. Anything that diversifies a one-crop cash economy, or that reduces Ghana's dependence upon a single crop like cocoa, the price of which fluctuates widely through circumstances over which the Ghana government has little or no control, is to be welcomed. If in addition it makes for more balanced regional development, particularly throughout the southern parts of the country, then it is an even more acceptable addition to the economy of a young and ambitious developing state in Western Africa such as Ghana has been throughout its period of independence.

PROBLEMS

Nevertheless there are problems, and it is important that they should be given careful consideration. Financially a development of this magnitude must be a great strain on a country for a long time. The tightening of the belt necessary now, with currency restrictions, import duties, various commodity shortages and the like, is something that Kwame Nkrumah, when President, often suggested was a temporary phase only. By 1970 he claimed that in Ghana there would be 'a State with a strong and virile economy, its agriculture and industry buoyant and prosperous; an industrialised nation serving the needs of its people' (*Plan*, p. xxii). 'No one,' he said 'will have any anxiety about the basic means of life, about work, food and shelter' (p. v). There would be no poverty and illiteracy, disease would be under control, and educational facilities would abound. Many years before—as long ago as 1953—he was saying, 'We are . . . entering upon this gigantic scheme, not as beggars, not as dupes, but as free and equal partners, the owners of our land and the masters of our own house'. But despite the optimism and confidence often expressed, it could be that relief will not come quite as soon as 1970: some observers indeed have been prophesying national bankruptcy for Ghana for years. It is clearly no easy task for a developing country to embark on a programme of this scale, no matter what political or financial support and interest it succeeds in attracting from outside.

There are also the problems of resettlement. Fortunately the lake area is very sparsely peopled and only 80,000 people are directly affected. But these 80,000 have to be settled on new land and supplied with the wherewithal of life, and the Government has had to embark on an extensive programme based on over eighty new sites, and with interesting experimental ideas for the development of new economies, involving the introduction of fisheries and the establishment of small-scale industries. In the long term, there are sure to be many advantages: in the short term; considerable expense (more than £98m.) is involved and some communities have suffered temporary hardship and frustration.

There are possible serious dangers to health. These have been very fully investigated by scientific and medical groups. There are dangers that the lake might be a routeway from north to south of disease as well as of people, and there is a possibility of an increase in bilharzia, through the breeding of the snails responsible in the lake. Very rigid control measures are, therefore, necessary. On the other hand, the risk of onchocersiasis or 'river blindness', which is transmitted by the *Simulium damnosum* fly, is lessened since *Simulium* breeds only in fairly swift-moving water, and the new lake will cover up many of its breeding places. The threat of river blindness is thus reduced by the creation of the lake, and the cutting off of the lower Volta from the source of the disease in the north should make it possible to wipe it out of the Ajena-Kpong district where at present the incidence is very high.

Other problems have already been mentioned in passing. They include the disruption of communications, as a result of the replacement of a relatively narrow river by a wide lake; the heavy cost of new communications, both the new ferries needed for the lake crossings and the fairly lengthy rail extensions that are essential for the exploitation of Ghana's supplies of bauxite; and the reshaping of the pattern of population and of labour supply in southern Ghana as a whole, and the temporary shortages of labour that might be experienced. Few of these problems need create difficulties of a lasting nature, and most of them can be solved quite quickly if they have been foreseen and the appropriate action considered at an early date.

In the long view, there can be no doubt that the Volta Project is an asset of tremendous importance for the people of Ghana, tight though the financial position and strained though the economy may be for the next few years. Here is a tremendous integrated programme, the ramifications and complications of which illustrate remarkably well the physical, financial, economic, social and political problems involved in any large-scale project in a developing country in tropical Africa today. The Volta Dam could indeed have significance for many other parts of Africa since there is little doubt that the statement made in the Preparatory Commission's Report, that 'failure or success of this scheme could have a profound influence on the possibility of raising finance for other large-scale schemes in the less developed parts of the world'

(*Report,* vol. 1, p. 2) is as true today as it was when it was first made more than ten years ago.

[1] The author is very grateful to Mr. D. Hilling, Lecturer in Geography, Bedford College, London (and formerly of the University of Ghana), who took him to the Volta Dam site and to the port of Tema in 1964, and to Mr. Stanton R. Smith of Kaiser Engineers and Constructors Inc., who supplied him with recent information and helped him in his understanding of some of the technical problems.

Figure 1 was drawn by Miss Janis Denton and Figure 2 by Mr. A. G. Hodgkiss, Senior Technical Officer of the Department of Geography, University of Liverpool.

[2] After the political changes of 1966 the Seven Year Development Plan was abandoned, and there has been considerable re-appraisal of many aspects of the Plan's proposals. The idea of lake ports was abandoned until further investigations had been carried out, and a transport survey was initiated. The Volta River Authority has recently opened a cargo service from Akosombo to Kete Krachi, the largest of the towns isolated by the formation of Lake Volta.

## BIBLIOGRAPHY


Food and Agricultural Organization of the United Nations, *Report of the Survey of the Lower Volta River Flood Plain,* Vol. 1, Rome, 1963.

*Ghana Seven Year Development Plan 1963/64 to 1969/70,* Accra, 1964.

W. A. Hance, *African Economic Development,* 1958, pp. 46–84.

........................., *The Geography of Modern Africa,* 1964.

D. Hilling, 'Ghana's Aluminium Industry', *Tijdschrift voor Economische en Social Geographie,* 1964, pp. 128–32.

T. E. Hilton, 'Akosombo Dam and the Volta River Project', *Geography,* 51, 1966, pp. 251–4.

Henry J. Kaiser Company, *Reassessment Report of the Volta River Project for the Government of Ghana,* California, 1959.

Kaiser Engineers and Constructors Inc., *Accra Plains Irrigation Feasibility Study,* I, II, III, Accra, 1965.

*The Volta River Project,* Vol. I, Report of the Preparatory Commission, Vol. II, Appendices to the Report of the Preparatory Commission, Vol. III, Engineering Report, Sir William Halcrow and Partners, H.M.S.O. for the Government of the United Kingdom and the Gold Coast, 1956.

Volta River Aluminium Scheme, Cmd. 8702, H.M.S.O., 1952.

CHAPTER V

# Legal Problems of the Volta Dam

L. ROUSE JONES

LAWYERS need to be brought into the negotiations for a Project of this kind at a time when by and large the lessons of geography have been learned, and by and large flexible economic scheming has been done. Before the lawyers joined the Volta negotiations it was known that the Volta dam, which was needed to develop Ghana, was to be made possible by selling about half the power to an aluminium smelting company backed by interests from overseas; leaving plenty of power, it was understood, for other purposes.

The moment for introducing lawyers is when an agreement in principle can be reached. The lawyers are expert at drawing out of the agreeement in principle the problems that lurk in it. They see aspects of their clients' interests that need protection or assertion, how proposed provisions might be vulnerable in certain events; and of course they take pains to make sure that they know how the agreements could be enforced. However, too much should not be put into the agreement in principle; unless it is clearly understood among the parties that much of it may need reconsideration before the agreements are finally signed.

In the legal work on the Volta Project, the things that caused the greatest amount of difficulty in negotiation arose from three basic considerations, as follows:

1. That one of the two main parties was the government of the country where the development was to take place.
2. That the other main party was a company owned by interests outside the territory of the government of the country. This factor is not an aspect of the first-mentioned factor, but is entirely different. Bound up with this consideration, however, is the fact that the second-mentioned main party was a company with limited liability.
3. The great cost of the Project in terms of money.

None of these three fundamental circumstances raised difficulties that the law reports or textbooks could help much to solve. But the documents involved the work of legal minds, and legal technique, and the making of a considerable complex of contracts, amounting to a body of quasi legislation to bind the parties.

The primary contract was a Power Contract under which future power was to be sold for at least thirty years to a Smelter Company at an agreed rate and with a minimum charge. The Smelter Company was backed by two American organizations, Kaisers and Reynolds. There were also loan agreements by which the Power Development and the Smelter were respectively financed. Loans were made by the World Bank, two Agencies of the United States government (the A.I.D. and the Export Import Bank of Washington) and to a smaller amount by the United Kingdom government. There were many other agreements containing inter alia undertakings to construct and operate the Power Development and the Smelter, and also, speaking broadly, to create the conditions and régime in which the parties were happy to come together. There were in fact twelve chief contracts, and a large amount of legislative, administrative and other documents, loan documentation, formal opinions of lawyers, in total about fifteen main documents and some 184 documents in all, for a Project planned to dispose of a possible total of £128 million.

I propose to discuss in turn the three fundamental considerations, and the way in which the problems they gave rise to were solved.

First, that one side was the government of the country where the dam was to be built. That side consisted of two organizations, the government and the Volta River Authority, a corporation that was to construct the Power Development through engineers and contractors, and then operate it.

The object of building the dam was to transform the country economically and industrially. It could not be built unless outside interests cooperated by taking about half the power for the purpose of aluminium smelting and other outside organizations provided large loans. It was not enough to fix the price of the power and other commercial terms as if the matter was an ordinary commercial contract within a country and leave it like that. An agreed background of legal rights and duties made unchangeable for thirty years and in some respects for perhaps fifty years had to be set up as well.

This background ranged from the special to the general. To start with the most special, it was agreed that the financial rewards to the parties were not to be left at large, and agreement was come to that the prices and charges fixed in the documents took account of fiscal charges, and that with certain exceptions no fiscal charges would be levied on the Smelter Company. This phrase was of course a very wide expression meaning taxation. For example there was to be no tax on the power used or on aluminium exported, and for example customs duties on articles to be imported by the Smelter Company were to be as agreed in the Documents. Of course income tax from the Smelter Company and its employees was one of the major rewards intended to accrue to the government; but it was agreed that it would be levied under the Ghana Income Tax Ordinance as it stood at the time of the Agreements, and a print of that Ordinance with amendments up to that time was bound up with the Project Documents. Agreement had to be reached for exemption from exchange control for the Smelter Company's construction funds and loan repayments, and for what nearly approached exemption for its operating funds, and the Smelter Company's funds were to be held outside the country. This concession as to holding its funds outside the country was secured for the Smelter Company by a method which was also used reciprocally to secure for the government important assurances of the Smelter Company's obligations. I will return to this matter.

The more general legal background needed by the Smelter Company was the ordinary legal background that would be taken for granted in the case of an ordinary commercial contract. For example the Smelter Company was to enjoy the usual borrowing powers of a company, the ordinary rights to water, and communications. It was also to have immigration rights for its expatriate employees, certain rights to acquire mining concessions and other matters. Indeed as part of the arrangements made for loans by United States Agencies the Agency for International Development stipulated that the Company should receive the most constant protection and security, words that suggest a trade agreement that might be entered into by a government for the benefit of its nationals in general.

The government of Ghana further undertook that there would be no derogation from the Agreements by legislation; no expropriation or nationalisation of or intervention in the Smelter Company or the Smelter for thirty years, and after that period had elapsed only on payment of full compensation; the provisions relating to this topic were developed in detail. There was to be no discrimination against the Smelter Company, its shareholders, creditors or employees; nothing which as it was put 'either formally or in true effect discriminates against' these persons, 'or which in effect, and looking at the matter fairly, singles out any of these persons'. It will be clear by now that much of this mass of undertakings amounted to undertakings against discrimination. Supplementary documents were to be provided by both sides as might be necessary. By statute especially passed, nothing in the main documents was to be illegal or void.

All this seems one-sided, but I will deal with the other sides shortly. In any case, these concessions whether of the special or the general kind were not drawn unnecessarily widely. Thus the Smelter Company's exemption from exchange control was very carefully limited to such amounts as the company could show were necessary, and necessary not merely for their purposes but for their purposes as agreed in the Documents. Other concessions as to exchange control were to meet certain obligations so far only as they were reasonable.

For a different reason, the fact that one of the two main parties was the government, and the government of an undeveloped country, caused problems in designing the organization that was to construct and operate the Power Development. It seemed obvious that the most suitable kind of organization was the statutory corporation, and the fact that the government would have what amounted to an equity investment in it did not seem to alter this view; for there would be no other equity investors in it. The kind of statutory corporation that manages a British nationalised industry was taken as model. But in an undeveloped state where governmental powers are concentrated at the centre, the question was could such a corporation function separately so far as day to day management was concerned; and could it be managed by industrial or commercial men, not by politicians on the one hand or civil servants on the other?

The question of providing for appropriate independence of action was faced. The President of Ghana assumed the chairmanship of the corporation. The President by the corporation's statute had power to determine the appointments of the other members at any time, except however, the appointment of the chief executive, who was to be appointed for a period. Subject to general control on matters of policy the chief executive was solely charged with the direction of the corporation's business and administration, and with the employment control and dismissal of all staff and workpeople. The appointment of chief executives was to be made with the approval of the Authority's chief lender, the World

Bank. A representative of the Smelter Company was to be appointed as one of two members of the corporation especially representing major consumers of power.

Thus this corporation, known as the Volta River Authority, was to be of a hybrid nature. For a statutory corporation to have the head of the government as chairman is no doubt not in character, but the chief executive was to have much of the function of the board in the model corporation, and the resulting compromise, or rather simplification, was probably the right one.

I now return to the second of the major considerations I have mentioned, i.e., that the Smelter Company was owned by corporations outside the country where the development was to take place; and was a company with limited liability. This was the other side of the matter. The problems it raised will take a little longer to discuss. That the company should have the backing of the corporations that were to set it up was of course the object from the beginning, and it was also thought that those corporations should assume considerable responsibility for its obligations.

Thus it was necessary to support the Power Contract by securing that the Smelter Company duly built the Smelter in accordance with its undertaking. This would support the obligation to pay for power and would also support the company's lender, and it was also an assurance for the Authority's lenders. Secondly it was necessary to have specific assurances that money for the Company's production would go to make the Company's payments.

Assurances were given. What form did they take? True to the genius of the Volta documents they were not given in the form of ordinary blanket guarantees. They were given in the form of two trusts. To secure that the Smelter would be built, the shares in the Smelter Company were transferred to trustees being a New York bank. If the Smelter Company became in default at any stage in regard to its obligations to build the Smelter, or in certain other events, then, if necessary after a special referee or ordinary arbitrator had decided that this was so, the trustee's duty would be, at the request of an interested party and with the consent of the Smelter Company's lender, to use the voting power of the shares to appoint an agreed manager to manage the company; and to appoint new directors to be nominated by that manager. Thus if construction according to plan was possible with a change of management, there was the legal machinery. The Smelter Company had the finance, so why should it not be possible? Presumably if *force majeure* intervened; and this trust also provided the way to test the question of *force majeure* at the right time.

If the Smelter could be built to time it would be, and if it could not be there would be no guarantors to sue.

The notion of this trust was that of the voting trust, well known in the United States as a device whereby shares in a company are transferred to a trustee for the purpose of using their voting power.

However, my statement in effect that construction of the Smelter should be possible needs further explanation. The voting trust was supported by a number of undertakings by the Smelter Company that it would, under any management, maintain its contracts prudently and commercially, and get in moneys to which it was entitled under them, and would not, for a further example, waive its rights to receive any of its resources. The Company had adequate resources. It had agreements with the corporations that owned it under which certain equity investments were to be made in the Company. The Company also had an agreement for a loan from the Export Import Bank of the United States; and if the Company received all these monies in the agreed manner and expended them on constructing the smelter and still had not enough then the Ghana Government was to make available the remaining amount of the cost by way of loan, an interesting further undertaking by that Government to cement the arrangements. Another New York bank was appointed a trustee with inter alia the duty to enforce payment of the Company's financing funds.

The Voting Trust Arrangements laid down the most detailed procedure whereby the fact of any default might be determined at or soon after the time when it took place, and whereby all interested parties might at that time have their say, and whereby steps might then be taken to put the matter right. Further, the Company was obliged to deliver certificates of progress to the Voting Trustees at certain stages, and also statements of days lost because of *force majeure*; and the interests owning the Company agreed to deliver statements of any default by the other side.

The difficulties guaranteed against in volumes of agreements have not so far been experienced in practice, and I was present in Ghana for the ceremony inaugurating Volta power early in 1966, the first generator having been switched on in September 1965 and the construction work was well up to schedule.

I have described in a few words the method of providing assurance that the Smelter Company would construct the smelter. By means of another trust more specific assurances were provided that the company would discharge its obligations to make payments for its power, and to make payments to its lender. This was the Current Accounts Trust, of which the Trustee is the other New York bank that I mentioned, in fact the Company's banker. Its banker was made trustee on behalf of the Authority and the lender to the Company and others to receive the moneys paid to the Company and to disburse them. It was to have power to enforce the payment to the Company of certain moneys, and I have already mentioned that this trustee was to supplement the powers of the Voting Trustee by enforcement of this kind. This trustee was to disburse the company's funds in stated priorities, as follows: after the trustee's own fees, first, to the Authority for power and to the Export Import Bank for principal and interest, and these two ranked *pari passu*. Secondly, taxation; but the government of the country had a certain right to require the trustee to call on certain credits so as to increase the funds at its disposal. Then in a stated order the trustee was to satisfy the other obligations of the company and to make payments to a special trust in Ghana set up by the corporations behind the Smelter Company to which trust a proportion of profits would be paid. Finally the trustee was to make payments to these corporations.

It may be pointed out that if the Company's lender in appropriate circumstances used a power under its agreement to make the entire principal of the loan due and payable immediately, then equal priority for the payment of the Authority's power charges would not be of much help to the Authority; and priorities tend to become important when something of the kind happens. It was directed that in such circumstances, or if damages were actually due and payable by the Company under the Power Contract, the trustee was to pay the Authority and the lender on the basis of average payments during the immediately preceding year.

Thus the same kind of solution was found to the problem of putting the Company in a position to pay its way during the operation period as had been found in respect of the problem of putting it in a position to construct the Smelter. The Power Contract provided for a power rate and a minimum charge for power and these had been agreed by all the interested parties; the owners had with similar agreement of all interested parties entered into agreements to pay for the Smelter's production at a rate of charge based on the world price of aluminium from time to time. The flow of payments was to be conducted by a trustee. How far did these arrangements give assurance that the Smelter Company would pay its way? They did not, and were not intended to, amount to a guarantee of the Company's commercial success; nor to such overwhelming assurances as were given for the construction of the Smelter: for the Company's charges for smelting alumina based on the world price of aluminium would not necessarily be enough in future circumstances to enable it to pay the agreed power charges. What the arrangements did secure in effect was that the Authority contracted directly with the corporations behind the Smelter Company, and not merely with their instrument the Smelter Company itself.

It was this Current Accounts Trust which made it possible for the Company's account to be held overseas.

I may mention that the corpus of Project Documents was flexibly drawn and it also contained guarantees in the usual form. Thus for example the government of Ghana guaranteed the Authority's performance of several of its agreements in a usual form.

There are many other safeguards in the Documents which support the system set up by these two trusts. Thus the agreement containing the undertakings between the government and the Smelter Company, called the 'Master Agreement', is not assignable without the consent of the other party to it. When the Voting Trust providing for the construction comes to an end transfers of the Smelter Company's shares will be permissible, and transfers of them subject to the Voting Trust are permissible now, but only between the two present owners or their subsidiaries, unless consent is obtained from the Authority. The Company is to be managed by managing agents but it can engage managing agents only in accordance with a special agreement. As I mentioned previously, the Company has undertaken to maintain various agreements in force notwithstanding any breach of condition by the other side and to act prudently and commercially under various agreements and enforce them and get in payments. The rights of its owners to make set-offs are restricted. The Company must not make payments to its owners if the payments would leave it too much indebted.

I may mention here that as well as the two main sides of the Project, there was in some sense a third side consisting of the various lenders, particularly as the Export Import Bank lent money to both the main parties. The common interest of lenders to the Authority was recognised in the Documents, and arrangements were made for some cooperation among them in administration and supervision of the loans. And the interests of many parties can arrange themselves in several patterns; an example of this is due to a coincidence in certain respects of the interests of the government and Authority on the one hand, and those of the lender to the Smelter Company on the other. This lender obtained undertakings from the Smelter Company that supported those which were obtained from it by the government or were undertakings that perhaps the government could not have obtained at all, e.g. as to maintaining its corporate existence, as to not merging with other companies, as to furnishing certain information, and many others. These were rights of the lender only, but any obligation on this Company of this sort was likely to be in the interest of the government as well. Another thing is that the mere fact of the payment out of the lender's money under its agreement would amount to assurance to the government that the conditions precedent to the loan had been fulfilled, including the conditions that the owners had put in their investments at agreed previous stages. The loan agreement gave the Authority power, if it chose, to cure a default in payment by the Company under the loan agreement, which was a right that the Authority might wish to exercise on terms, so as to keep the Project in existence.

I have now described and discussed the problems caused by the first two of the fundamental considerations I referred to, except the remaining problems of enforcement, and the solutions found for these problems. I turn now to the other major consideration, the large amount of money involved. These amounts were to be possibly £70 million for the Power Development and possibly £58 million for the Smelter. The Power Development has been built now and has not cost so much. In addition, the government had spent large sums on a new port and roads, partly for the purposes of the Project.

I think that the large amounts of money involved caused two groups of legal problems only; first, problems in regard to fixing future dates having regard to possible *force majeure* and other uncertainties, for if there were serious delays additional interest charges and other costs would add big sums to the cost of the Project; and secondly problems caused by the bringing in of a large number of parties which had to be brought in because of the large amounts involved.

The most important future dates were:

1. the date on which the permanent delivery of power to the Smelter was to start,
2. the date when the minimum charge to the Company was to begin, and

3. the dates by which the Smelter Company undertook that the Smelter was to be built, a date not necessarily the same as the date by which they intended to have it built if possible.

The permanent delivery date was to be fixed within limits by the Authority. But the Authority had a certain power with the consent of its lenders to postpone the later limit in its direction on giving reasons, obtaining the consent of its lender, and paying to the Company certain out-of-pocket expenses. The Authority was to have additional limited power to postpone the date because of any *force majeure* that might arise.

The minimum charge was to be payable from the permanent delivery date even if *force majeure* was acting on the Company, unless the Company had started construction of the Smelter early and had continued it with reasonable diligence or had tried to do so and then *force majeure* was preventing the using of the power.

At the permanent delivery date there was in the absence of a special notice no obligation to have the Smelter ready, the Company being obliged merely to pay for the power and to have the Smelter ready by the second anniversary of the permanent delivery date, with a certain rated capacity, but the Authority could oblige the company to have the Smelter ready at the permanent delivery date, though with a much less rated capacity, by giving a longer notice, on pain of paying out-of-pocket expenses if it failed to deliver power; these expenses would have to be paid even if its failure was caused by *force majeure*.

I have now stated what is the only instance in the Documents where a party was to be liable to pay for something though prevented by *force majeure* from enjoying what it was paying for, and what is the only instance of a party being liable in what amounts to damages, albeit limited damages, for a failing caused by *force majeure*.

An odd position arose. If the Authority gave the ordinary, not the special, notice to fix the permanent delivery date it would not be liable for non-delivery of power if its failure was caused by *force majeure*. Nevertheless, however *force majeure* might strike, the Authority was to have no indefinite power to postpone the permanent delivery date. Thus there might be a permanent delivery date willy nilly at last declared in spite of *force majeure*, but with the Authority still producing no power because of *force majeure*, and the Authority not liable. This might be merely consistent but it would not be sensible, and in such circumstances, the Smelter Company would have special power with the consent of its lender to rescind the Contracts.

There was a reciprocal arrangement. If the Smelter Company on its side under the special provision I have mentioned started the work of construction of the Smelter by a certain time and continued with reasonable diligence, then if *force majeure* supervened it would be excused from paying the minimum charge. But the delay in paying the minimum charge was not to exceed three years: it is expressed that if it did exceed three years, the Authority would have power with the consent of the same lender to rescind the Contracts.

It will be obvious that the arrangements for fixing these three future dates were arrived at after much thought and discussion. It may perhaps be thought that the resulting provisions were too much developed, too complicated, and too detailed. The account I am able to give here spares you a sight of the detail, but I assure you that it is measured in pages rather than words. It may perhaps be thought also that the element of reciprocation, of mutality, is overdone. Such criticisms may and may not be well-founded, but if they are they are not relevant when the Documents are considered as precedents or forms for help in framing the documents for other projects: because the complications and details make it possible for others to see more clearly what they might leave out, by giving them a clearer idea of what the position would be if all these things were put in.

The arrangements I have just outlined also show what a large part the possibility of *force majeure*, normally a very minor matter in a contract, played in complicating the Documents. This was only partly because of the large amounts of money involved. The other considerations give rise to special risks of *force majeure*. *Force majeure* was very widely defined; everything was included, even shortage of water. The common requirement was that the result of the force designated as *force majeure* must be unavoidable by proper precautions or reasonable alternative measures. The government gave undertakings that certain acts that might otherwise amount to *force majeure* would not occur, but *force majeure* as defined for purposes of the Project was very much wider in scope than matters covered by those undertakings. The three kinds of *force majeure* were all included in full: violence such as war or mob violence, natural disaster such as epidemic or earthquake, and governmental action such as licensing or embargo. It should be noted that a relevant war might be a war anywhere in the world, and that governmental action of the kind intended might be the action of any government.

If *force majeure* caused damage or destruction to the Power Development or the Smelter, the Authority or the Company as the case might be was to reinstate it so far as necessary. However, alternative facilities could be set up instead; Volta power is cheap but at some time nuclear power may be cheaper. There was to be no obligation to reinstate if the cost of doing so would exceed one-tenth of the original cost of construction: in such circumstances the parties were simply to consider what should be done.

*Force majeure* was to suspend a party's obligations while it continued, and if not cured it might bring the Project to an end. The matter leads me to my final topic, enforcement, for breach and *force majeure* are related topics in the Documents.

The discussion of *force majeure* has been in part a digression from the topic of the problems caused by the large cost of the project. In addition to causing difficulty over future dates, the basic fact of this high cost gave rise to a multiplicity of parties, and therefore of contracts. This phenomenon led in turn to complicated methods of knitting the complex of contracts into what was broadly speaking, and in essence, one single many-sided contract. It may be objected that even if the Project had cost much less it would still have been multi-partite, so that this was a basic consideration of its own.

Thus there arose many third party rights. I have already described the two trusts, under which all interested parties were beneficiaries. I have mentioned also the Smelter Company's contractual obligations to the government and the Authority to maintain its financing and other agreements in force even if circumstances arose where at common law it would have been entitled to rescind for breach of condition, and more positively to act under its agreements and get in moneys to which it was entitled under them, and not waive its rights. This kind of provision gave third parties rights of a contractual nature. The Authority entered into a similar obligation to one of its lenders. One way and another many third party rights were distributed to the various parties. Another method was to require the consent of one party to an act of another.

The existence of many third party rights did not lead to any restriction of the parties' rights to enforce the Contracts by claiming damages; but they, except in the loan agreements, did make necessary the substitution for the common law right of rescission for breach of condition of an immediate right to suspend its own performance; only in certain expressed circumstances could the government side or the aluminium side determine its obligations.

One kind of party, the loan creditors, had wide powers to terminate at their discretion, or to suspend instead. This is of the nature of banker's agreements, and no doubt all loan agreements stop if financial unsoundness appears in the borrower. In these agreements the power to terminate took the form of power to declare the principal debt due and payable immediately and it was conferred by the loan in case of breach and in case of certain kinds of *force majeure* indifferently. Indeed it might not in the event be possible to distinguish between the two. For example, assimilation of breach and *force majeure* arose in a provision allowing the lenders to terminate or suspend if what was described as an 'extraordinary situation' arose which made it 'improbable' that the borrower or its guarantor would be able to perform its obligations to the lenders. Speaking broadly the Authority's loans could be called in in this way if construction or operation did not proceed or if there was any breach of the obligations on the government side that I described earlier; and indeed the Authority's loan could be called in if the Smelter Company's loan was called in—just for that reason. In the case of the Smelter Company's loan, any breach of one of the Project contracts to which the Smelter Company was a party, even if a breach by the other side, would entitle the lender to call in the Company's loan. Such contingencies cannot be described so far as the innocent party is concerned as breach, or as *force majeure*; they can be described in no more specific way than as conditions subsequent. But of course it had to be accepted that if termination of one contract occurred it might well bring down all the others with it.

There is no time to give even a sketchy general account of these powers to terminate and suspend as they appear in the corpus of Documents. Many are aimed against expropriation or similar action. They may perhaps seem drastic powers, but they are not more so than the ordinary common law power to rescind for breach of condition, indeed they may be less so because they are restricted so as to make so far as possible for the Contracts going on to complete performance rather than being called off. They occupy many pages of the Project Documents, but express powers of enforcement had to be subsitituted for what the common law would have inferred. The fact of their expression makes them look like statutory penalties, but that is not their nature. Powers to enforce a contract are the force that binds a contract together, and are an essential part of a contract.

One has to think how the various interests would act in circumstances that might arise. The loan creditors for example would be likely, if the question arose of calling in their money, to find themselves with large amounts outstanding in the Project, and would need to have power to exercise influence on the decision what was to be done. A party's rights to suspend or terminate was restricted to its rock-bottom requirements. Thus the Authority and the Company each had ultimate power to rescind the Power Contract, but only ultimate power.

These powers of enforcement were probably provided in these Contracts in such form as would achieve their purpose. But it would be an enlightening and valuable work, also a fascinating and an ingenious one, to frame provisions for some future project that would provide for the consequences of breach and *force majeure* in the simplest possible terms, and without repetition. It would make possible a much clearer grasp of this aspect of the matter.

CHAPTER VI

# The Impact of the Owen Falls Hydro-Electric Project on the Economy of Uganda[1]

WALTER ELKAN and GAIL G. WILSON

NOTHING gives rise to so much general satisfaction as a new electricity station. Electricity is regarded as the hallmark of a modern society, providing people with light and heat in the cleanest and most convenient form. It is also deemed by many to have overriding importance for a country anxious to accelerate economic development. The Owen Falls Hydro-Electric Project in Uganda was especially intended to promote the latter and this paper is therefore concerned with two question:—first, what part hydro-electricity was expected to play in the economic development of Uganda and, secondly, whether ten years of operation have born out these expectations.

In contrast with for instance the Volta River Scheme, the Owen Falls Dam was not designed as a multi-purpose project; its principal purpose was to provide a supply of hydro-electricity for Uganda[2]. It is built across the River Nile at Jinja only a few miles from Lake Victoria and the Lake acts as a natural reservoir. The dam is 2,500 feet long and 100 feet high. It controls an average flow of 42,000 cusecs and the full capacity (ten generating sets) will be 150 MW. The generator sets are small by international standards because the railway from Uganda to Mombasa was unable to carry anything larger.

Those who are familiar with the folklore of Uganda know that the Owen Falls Dam is a lovely dream come true. To quote Kenneth Ingham, it had 'both fascinated and unnerved administrators for half a century'[3]—in fact ever since Churchill had given them the idea after his African journey in 1905[4]. But the inter-war years were not propitious for a project of this magnitude. Indeed, with the exception of the railway extension into Uganda there was not a single item of capital formation between the two world wars on which even £0·5 million was expended. This is hardly surprising when one recalls that government revenue never exceeded £2 million in any year and that the generally low level of effective demand made capital investment in industries serving the home markets unattractive. Uganda grew food, exported £3·5 million (1938) worth of cotton and a small amount of coffee and imported most of its other needs. 'Industry' was virtually confined to the cotton ginneries, a few other agricultural processing concerns and three cigarette factories. No wonder then that a survey of the possibility of generating hydro-electricity from the Nile, made in 1935, concluded that such a project, though technically feasible, would not pay. Actual and potential electricity consumption were judged to be too low[5].

The history of electricity in Uganda is to some extent the history of the government's changing view of its economic role. Before the war it had seen as its main task the expansion of health and education services as far as revenue permitted. When revenue began to increase during the war, the first impulse was to raise expenditure on these two services, and indeed, the 1944 Development Plan was largely concerned with how to do just that. However the Plan proved unacceptable to the newly appointed Governor, Sir John Hall, whose notions about development were at variance with his predecessor's. He commissioned a distinguished natural scientist, Dr. E. B. Worthington, to draw up a new plan. His introduction to that plan laid the foundation of post-war policy: —

> 'That a great expansion of health and education services is eminently desirable is beyond question, . . . but meanwhile the revenue for that expansion does not exist and . . . our only course is to press on with the production of national wealth and to restrict the expansion of social services within the limits imposed by the funds that can be made available'[6].

This doctrine of the primacy of economic development inspired all the major capital projects in the decade following the end of the war, including the Owen Falls Dam.

The change in the government's view of its functions coincided with a dramatic change in the economic possibilities that were open to Uganda. The value of exports which had been under £5 million in 1938, was £10 million in 1945 and £50 million seven years later. Government revenue was £2 million in 1938, £3·5 million in 1945 and £16 million in 1952.

The new Plan therefore had far more funds at its disposal and was soon to have even more. Two factors explain why the Worthington Plan has sometimes been taken to have placed unique emphasis on the development of agriculture, as distinct from industry. They are, first, the importance of peasant farming in a country of five million inhabitants of whom only some 250,000 worked for wages and secondly, the limitation of the Plan to what were deemed the proper objects of government spending. Consequently, the lack of a section on industry must not be taken as indicating that industrial development was not envisaged. On the contrary, the Governor's foreword made it clear that he gave high priority to industrial growth but that the means to attain it was not through the Development Plan. A fundamental pre-requisite was held to be the expansion in the supply of power, for how could industry develop when the total public supply of power was already committed. In 1945 power generation was limited to a small wood-burning station which supplied Kampala and Entebbe and a diesel station for Jinja. Some industrialists and missions generated their own electricity from imported fuel, but electricity was definitely a luxury. Industrial development was deemed to need 'ample cheap power'. This catch-phrase came to be repeated over and over again.

The Governor therefore arranged for Mr. (later Sir) Charles Westlake to make what amounted to a feasibility and market survey of the Owen Falls Project. His report[7] was accepted in 1947 and implemented with commendable speed. The Uganda Electricity Board was set up as independent public corporation with its finances guaranteed by the government, and work on the site of the dam began in 1948. Despite some difficulties and many delays, the first two generators came into operation early in 1954.

We now turn to the impact that hydro-electricity has had on the economy. Table I sets out the basic facts of electricity supply and demand over the years 1954 to 1965.

The table shows that installed capacity was raised quickly to 120 MW by 1958, leaving two more generators to be installed before full capacity of 150 MW is reached. When, for reasons to be examined later, consumption increased less rapidly at first than had been hoped, the Uganda Electricity Board entered into a 50 year bulk contract with Kenya to supply up to 190 million units at a rate very favourable to Kenya. The table shows how prominently the Kenya 'off-take' has figured in the total units sold, never accounting for less than one third of sales and sometimes for almost one half. The Kenya 'off-take' had the effect

TABLE I

GROWTH OF ELECTRICITY CONSUMPTION 1954–1965

| *Year* | *Installed Capacity MWh* | *Total Sales Million kWh* | *Proportion of Total Sold to Kenya %* | *Maximum demand as % of Installed Capacity* |
|---|---|---|---|---|
| 1954 | 30 | 61 | — | 50 |
| 1955 | 60 | 69 | — | 26 |
| 1956 | 60 | 82 | — | 36 |
| 1957 | 90 | 134 | — | 33 |
| 1958 | 120 | 253 | 36 | 49 |
| 1959 | 120 | 315 | 41 | 48 |
| 1960 | 120 | 363 | 44 | 52 |
| 1961 | 120 | 400 | 48 | 61 |
| 1962 | 120 | 417 | 45 | 61 |
| 1963 | 120 | 461 | 41 | 71 |
| 1964 | 120 | 471 | 38 | 83 |
| 1965 | 120 | 522 | 36 | n/a |

Source: Uganda Electricity Board, *Annual Reports*

of making the increase in the number of units sold in each year correspond fairly closely to the original forecasts on which the scheme was based[8], but maximum demand rose much more slowly and consequently the power house did not even begin to approach installed capacity until 1963.

Electricity was used but not by the types of consumer or in the places that had been expected. Small scattered consumers predominated where large factories or plants concentrated in Jinja and Kampala had been expected. One reason why consumption did not measure up to expectations was that although power was available, it was not cheap for the majority of consumers. This in turn was because the costs of generating turned out to be much higher than had been anticipated. In electricity generation cost is largely determined by capital repayments and interest on outstanding debt. These proved to be much bigger than expected. In the first place the project suffered from the general post-war rise in prices of imported materials and equipment. The *Uganda Electricity Survey* had put the capital cost of the dam, generating station and transmission and distribution system at £4·25 million in 1947. A year later the 'project report' raised the figure to £7 million, but the final cost has been in the order of £30 million, with £16·5 millions for the dam and power station alone. In addition, recurrent costs have been raised by the rise in the general level of interest rates on loans and the situation was exacerbated by a policy of borrowing on fairly short-term loans for part of the finance. Westlake in the *Uganda Electricity Survey,* had warned that every increase of 1 per cent in the average rate of interest (over the expected rate of 3 per cent) would result in a 10 per cent increase in the cost of producing electricity. The average rate of interest on the Uganda Electricity Board loans rose above the critical level of 3 per cent in 1952 and to 5 per cent in 1962.

It also became necessary to replan the transmission and distribution system. Westlake had envisaged a system which would supply industry and the more affluent householders in Kampala, Jinja and Entebbe, and at a later date Masaka. The system would therefore be short and relatively compact. However, when the Uganda Electricity Board adopted a policy of selling electricity virtually wherever a consumer could be found, a much more extensive network had to be built. Distribution costs rose and at the same time the system became more vulnerable to certain adverse features of the Uganda environment. The area has one of the highest liabilities to thunderstorms in the world and experience soon showed that rural lines and small substations were particularly vulnerable to damage by lightning and extra protection had to be installed. The climatic problems are particularly severe in the south of Uganda where the network is most extensive and where moreover the countryside also presents most difficulties. Lines had to be built across swamps, through banana groves or in dense secondary forest. Compensation for damage to crops reached a peak of more than £50,000 p.a. in 1959 and lines through forests have to be frequently checked to make sure that creepers do not climb the poles and cause short circuits.

The pattern of settlement in Uganda presents a further problem. Outside the main towns, the great majority of the population live on isolated farmsteads and not in nucleated villages. Even the rural industries, cotton ginneries and coffee hulleries, have often been located singly without a surrounding village, and trading centres are small and with few exceptions, poorly developed. This makes the cost of extending supply to the rural areas much greater because there is little chance of picking up extra revenue at intermediate points along a line. It is estimated that in Great Britain the capital cost of connecting a new consumer is under £100; in Uganda it is £250 and steadily rising as the distance from the main power station gets ever longer. Although over three-quarters of all electricity is in fact sold in the towns, the Uganda Electricity Board has pushed the network far out into rural areas, partly for social reasons, but mainly because it urgently needed to sell electricity and was willing to supply isolated factories and trading centres even at a high cost. It is therefore not surprising that out of the total capital invested in the generation, transmission and distribution of hydro-electricity, only 55 per cent is accounted for by the dam and power station: all but 5 per cent of the remainder has been the cost of creating the distribution network.

The following table displays total sales of electricity by category of consumer. (Due to the omission of certain minor classes of consumer the totals do not add up to 100 per cent). The significance of the distinction between 'ordinary' and 'special' industries is explained below.

TABLE II

ELECTRICITY SALES TO DIFFERENT CATEGORIES OF CONSUMER

| *Year* | *Domestic* | *Commercial* | *Industry (Ordinary)* | *Industry (Special)* | *Kenya off-take* |
|---|---|---|---|---|---|
| | % | % | % | % | % |
| 1954 | 33 | 6 | 48 | — | — |
| 1955 | 36 | 7 | 43 | — | — |
| 1956 | 37 | 7 | 43 | — | — |
| 1957 | 26 | 5 | 23 | 37 | — |
| 1958 | 15 | 3 | 14 | 27 | 36 |
| 1959 | 13 | 3 | 13 | 26 | 41 |
| 1960 | 11 | 4 | 12 | 25 | 44 |
| 1961 | 10 | 4 | 12 | 23 | 48 |
| 1962 | 9 | 4 | 12 | 26 | 45 |
| 1963 | 9 | 4 | 14 | 29 | 41 |
| 1964 | 8 | 4 | 14 | 33 | 38 |

Source: Uganda Electricity Board, *Annual Reports*

Table II shows how the relative importance of domestic and industrial consumers has changed over the years. As the output of electricity increased the percentage used by small consumers, (domestic and commercial), fell, while the proportion for the two classes of industry remained relatively steady. As was seen in Table I, an increasing share of the output was exported to Kenya.

Although the relative importance of domestic consumers has declined, Table III shows that the total number of consumers rose each year, and until 1961 so did the total number of units used. Table III illustrates this growth, but it also shows that the average number of units used per consumer has fallen by over 50 per cent since 1954. Although some of the decline in consumption per head may be attributable to the rise in charges, (domestic tariffs were raised in 1959 and again in 1962), the major reason has been the decline in the proportion of non-African consumers. This is strikingly shown by the fall of total consumption in 1961, which was clearly the result of a decline in the number of European and Asian consumers as independence approached. Few Africans and Asians can afford to use electricity as lavishly as expatriates, and even when they can and have moved into houses and jobs formerly occupied by Europeans, there remain cultural reasons why they use less electricity. For instance, most Africans prefer the South Uganda staple of bananas to be cooked over charcoal and the Uganda Electricity Board's enterprising attempt to popularise a

Table III

DOMESTIC ELECTRICITY CONSUMPTION

| *Year* | *Number of Consumers* | *Units used '000 kWh* | *Average kWh/Consumer* | *Average Revenue per unit cents* |
|---|---|---|---|---|
| 1954 | 5,760 | 21,362 | 3,700 | 20 |
| 1955 | 7,220 | 25,384 | 3,520 | 18 |
| 1956 | 8,560 | 30,436 | 3,560 | 18 |
| 1957 | 9,970 | 34,971 | 3,510 | 18 |
| 1958 | 11,620 | 38,912 | 3,350 | 18 |
| 1959 | 16,120 | 40,795 | 2,530 | 22 |
| 1960 | 17,850 | 41,020 | 2,300 | 24 |
| 1961 | 19,240 | 39,638 | 2,060 | 27 |
| 1962 | 20,530 | 38,878 | 1,900 | 30 |
| 1963 | 23,000 | 40,456 | 1,760 | 30 |
| 1964 | 24,190 | 40,101 | 1,660 | 31 |

Source: Uganda Electricity Board, *Annual Reports*

specially designed banana steamer has met with little success. In 1962 European domestic consumers in Uganda were using ten times as much electricity, on average, as Africans and Asians three times as much. Hence as more Africans and Asians took electricity the average per consumer fell.

Domestic electricity is expensive. The minimum charge is sh. 8/50 (8s. 6d.) a month, and the average price paid by domestic users has risen from 18 cents per unit in the late 'fifties to 31 cents in 1964. After ten years of ample electricity, there were only 23,000 households using it out of a population of nearly seven millions. To quote David Walker, hydro-electric power 'has made life rather more pleasant for the European population and the better-off Asians and Africans in the part of the country that the power lines reached, but it has not made a big difference to them; and the mass of the population has been entirely unaffected'[9].

The same is true today, and is likely to remain so for some time to come.

It might be suggested that a different system of tariffs would have resulted in greater benefits for Uganda as a whole. This is, however, extremely doubtful in view of the constraints imposed on the UEB by circumstances outside its control. In the first place Kenya and the Special Industries have consistently used 70 per cent of the electricity produced during the last five years. The UEB could not have charged them a significantly higher price. The price of power sold to Kenya had to be competitive with the cost of electricity generated from imported fuel oil in 1958 and Kenya was for many years unwilling to renegotiate the agreement before it expired. The tariffs for the Special Industries had to be low enough to attract and keep the overseas investors concerned in competition with possible locations in other countries. There was therefore virtually no possibility of raising charges to the larger consumers so that small users could benefit.

It is conceivable that tariffs for ordinary industrial power might have been lowered and the charges raised for the smallest consumers. However a fall in industrial tariffs which was large enough to act as a stimulus to industrial development, and financed by higher charges for non-industrial users, would certainly have led to a fall in consumption and revenue among the latter. In 1960, this non-industrial group contributed 52 per cent of total revenue and used 19 per cent of the units produced. For this group electricity was a luxury good and any further rise in price would almost certainly have led to a fall in total revenue and there is no reason to believe that the change would have been followed by an increase in industrial consumption great enough to justify the measure.

The situation in relation to differential charges according to location is very similar. Approximately 75 per cent of the electricity sold in Uganda in 1960 was used in Kampala and Jinja. Higher charges in the outlying areas would not have brought in sufficient revenue (assuming consumption levels were unchanged) to justify a significant drop in price in the main centres. Neither would differential revenues cause a greater concentration of economic activity in the towns. The main outlying consumers were either in smaller towns, or were tied to their raw materials (cement, sugar and tea), or were enabled to continue in business by favourable legislation (cotton and coffee factories), or were unlikely to move for other reasons (missions). Under these circumstances higher electricity charges would clearly lead to substitution of other forms of power, increased surplus capacity at Owen Falls and a lower return on investment for the scheme as a whole. The UEB has therefore had very little room for manoeuvre in formulating its tariffs and only a heavy government subsidy could have enabled it to depart significantly from the framework adopted. The growth of industrial consumption in the future would, however, allow more flexibility.

The Owen Falls Dam was built primarily as an inducement to the development of industry. The excess capacity revealed in Table I, especially in the early years of operation, is one indication that industrial development did not take place on the scale envisaged. Table II shows that until the post-independence phase of expansion began in 1964, industrial consumption in Uganda rarely accounted for more than about 40 per cent of the Owen Falls output whilst the Kenya offtake was never less. It must also be remembered that 'industry' is defined in terms of electricity tariffs and some of the consumers who take electricity on an 'industrial' tariff would not normally be classed as industries; many more are services rather than manufacturing concerns. For example in the Kampala area hospitals, waterworks and laboratories together used approximately half of the 'industrial power' sold in 1960. Since 1963 the largest hospital in the country has counted as a 'special industry'. There are also a few cases of industries using non-industrial tariffs, the chief being a biscuit factory and the main railway workshops. The fact that it is worth mentioning them indicates the small size of the industrial sector. Table II divides industrial consumption between Ordinary and Special industries. The Special industrial tariffs are individually negotiated for each consumer and, in 1965, only applied to the six largest customers. They were the textile mill, the copper smelter and the steel works at Jinja, the copper mine in Western Uganda, the cement works and the hospital which has already been mentioned. Other industries paid between three and four times as much for power as these consumers. In 1964 the 'special' industries accounted for 50 per cent of Uganda sales of electricity and for 13 per cent of the UEB's total revenue.

Industrial development was not included in the Uganda Development Plan of 1947. At that time it was confidently expected that the various industries which were making tentative enquiries would quickly establish themselves in Uganda. This did not happen. It was some years before the government recognised that new industries would not flock spontaneously to Uganda, and that active steps were required to promote industrial development. An important step was taken when the Uganda Development Corporation was set up in 1952. It was asked to take over existing government industries, endowed with £5m capital and told to make itself responsible for sponsoring likely projects in partnership with private firms especially from overseas. Its capital was to be used to entice reluctant overseas businessmen: if a quasi-government corporation was prepared to shoulder part of the risk and participate in a venture and also provide local knowledge, an overseas business might be more willing to invest some capital and above all impart some of its industrial expertise.

Another step taken to promote industrial development was to create an industrial estate in Jinja. Railway sidings were laid out and an improved network of access roads built. While work on the dam continued all this seemed sensible. The labour force in Jinja reached a peak of 2,500 Africans, 200 Europeans and 123 Asians. Many of them brought their dependents and as would be expected, the sudden increase in population brought new prosperity to Jinja. Offices, shops and flats were built and an industrial future confidently predicted. However, once the dam was completed in 1954, the labour force fell rapidly and stood at only 600 by the end of 1955. Jinja was left with empty buildings and decaying streets. Site after site on the industrial estate had to be let to firms requiring storage space rather than room to manufacture. Population and economic activity were higher than they had been in 1948, but the progress was obscured by the fall from the 1953 peak and an atmosphere of depression and slump pervaded local business circles[10].

The Uganda Development Corporation was not at first successful in its efforts to attract industry from overseas. There was a gap of nearly two years between the time when hydro-electric power became available and the first major customer took a supply. This was the textile mill which began production in 1956. A copper smelter, also in Jinja started up the following year. Meanwhile the cement factory in Tororo, built to supply the Owen Falls Dam with locally produced cement, was actually finding it cheaper to import wattle wood from Kenya and use charcoal for most of its power needs. Agreement on the bulk supply to Kenya was reached in 1957 and export began in 1958. The UEB was thus assured of three major customers and a market for a substantial proportion of its output, but there was no cause for complacency. The textile mill experienced financial difficulties which were not completely overcome until 1961 and the UEB would have been in a very difficult position if the mill had closed. In fact the Board was only able to show a working profit by charging part of the interest due on past borrowing to the capital account. 1964 was, in fact, the first year in which the UEB made a surplus after covering all costs that are properly charged to revenue. In 1961 and 1962 the accounts had shown annual deficits of £350,000.

Industrial development was not confined to a few large concerns, even if 'large' in this context is relative to Uganda alone. Perhaps equally significant from the point of view of economic development was the mushroom growth during the 1950s of small concerns serving the expanding home market. The back-yard furniture work, primitive soap factories, bottling plants, printing presses, bakeries and biscuit factories are 'industries', and it has yet to be demonstrated that concentration on large ventures is necessarily the quickest way to development. Small concerns at any rate tend to be less capital intensive and more manageable in a country where few people have had experience of starting and running large undertakings.

These small new industries, and others such as the expanding activities of Uganda's two sugar millionaires, together with the industries which existed before hydro-electricity became available, constitute the bulk of the small industrial consumers. They developed to some extent in all the towns but the main growth was in Kampala where income per head and total income per square mile provided the biggest market for consumer goods. They were not power oriented, and in any case since the UEB charged a uniform tariff throughout the country, there was no point in locating themselves near the source of supply. Until quite recently, industrial development in Jinja lagged behind Kampala. It may well be that most of these concerns would have come into being even if the hydro-electric project had not been undertaken and the rate of industrial development has in any case been low. Table IV gives some indication of the role of industry in the development of Uganda since 1954.

It shows that the contribution of manufacturing to total national income has remained adamantly at about 5 per cent and the number employed in manufacturing industry has remained steady at some 25,000, accounting for 10-12 per cent of a total labour force which has slightly declined. Electricity was available in ample supply for new industries, but for most consumers it was not cheap. Table V, which shows the average revenue per unit sold to different classes of consumers, gives some indication of the cost of electricity.

TABLE IV

ELECTRICITY AND MANUFACTURING INDUSTRY

1954 – 1963

| (a) | (b) | (c) | | (d) | (e) |
|---|---|---|---|---|---|
| *Year* | *Total units sold in Uganda million kWh* | *Gross Domestic Product*[1] *£m* | *% of GDP generated by Manufacturing*[3] | *Total African Employees in Enumerated Concerns*[4] *thousands* | *African Employees in Manufacturing*[3] *thousands* |
| 1954 | 64 | 129 | 4 | 225 | 24 |
| 1955 | 69 | 140 | 5 | 226 | 26 |
| 1956 | 82 | 142 | 5 | 225 | 25 |
| 1957 | 136 | 147 | 4 | 227 | 25 |
| 1958 | 163 | 147 | 4 | 228 | 25 |
| 1959 | 185 | 149 | 4 | 224 | 24 |
| 1960 | 202 | 152 | 4 | 229 | 25 |
| 1961 | 209 | 156 | 4 | 221 | 26 |
| 1962 | 228 | 157 | 4 | 217 | 25 |
| 1963 | 271 | 176[2] | 4 | 208 | 25 |

NOTES:

1. This included subsistence production which was estimated to have been £36 million in 1954 and £47 million in 1963.
2. Estimated. About £11 million of the increase between 1962 and 1963 is accounted for by an increase in the output of agriculture.
3. This does *not* include crop processing, whereas 'Other industrial consumers' in Table II does.
4. Excludes an estimated 60,000 working for African farmers, mainly in Buganda, and all others working in concerns which employ less than 5 persons.

Sources:

*IBRD Economic Development of Uganda* Table II
*IBRD Report on the Current Economic Position & Prospects of Kenya Uganda & Tanganyika* in Table III (a) (Uganda)
*Uganda Statistical Absract* 1959 and 1964

TABLE V

AVERAGE REVENUE PER UNIT SOLD 1954 – 1964

| | *Domestic cents* | *Special industrial cents* | *Other industrial cents* | *Kenya Bulk cents* |
|---|---|---|---|---|
| 1954 | 20 | | 19 | — |
| 1955 | 18 | | 18 | — |
| 1956 | 18 | | 15 | — |
| 1957 | 18 | 7 | 18 | — |
| 1958 | 18 | 6 | 17 | 3 |
| 1959 | 22 | 6 | 18 | 3 |
| 1960 | 24 | 6 | 19 | 3 |
| 1961 | 27 | 5 | 20 | 3 |
| 1962 | 30 | 6 | 22 | 3 |
| 1963 | 30 | 5 | 21 | 4 |
| 1964 | 31 | 5 | 22 | 5 |

Source: Uganda Electricity Board, *Annual Reports*

The 'special industries' that took over half the electricity sold in Uganda in recent years, got their power as cheaply as any one could have hoped (approximately 5 cents per unit or 0.72d). The average paid by other industrial consumers, about four times as much in the last four years, is not prohibitive, but neither is it low enough to stimulate industrial development when other circumstances do not favour it. This is merely a variant on the theme with which Professor Hirschman's work has familiarised us, that investment in Social Overhead Capital, like a hydro-electric project, cannot in itself be depended upon to stimulate industrial development[11].

Spontaneous industrial development has been mostly on a small scale and much more the result of the buoyant local demand generated by rapidly rising farm incomes in the 1950s than a consequence of the availability of electricity. The few bigger enterprises which came into being during this period owe their existence to the active endeavours of the public authorities in interesting overseas firms. The promise of ample electricity at low rates, however, certainly played an important part in the setting up of the first textile mill and the copper industry.

When the decision to build the Owen Falls Dam was taken, it was hoped that the greatly enhanced supply of electricity would be almost all that was necessary to bring about a rapid development of secondary industries in Uganda. The press and publicity services talked in terms of Jinja as 'the Detroit of central Africa', while the Worthington Plan said that 'experience with electricity in other parts of the world has nearly always shown that the most optimistic estimates of consumption have been greatly exceeded soon after the provision of a reliable and cheap supply'[12]. It was expected that cheap and ample electricity would attract many overseas investors anxious to set up large manufacturing concerns. It was also taken as automatic that this would be very desirable since it would help to raise the level of living.

These expectations cannot be said to have been fulfilled. The extent of overseas investment has been very small and its contribution to raising the level of living has been marginal if indeed there has been such a contribution at all. To take only one example:—it is very doubtful whether the Nyanza Textile Factory in Jinja would ever have overcome its initial difficulties if the import duty on all textiles had not been substantially raised in 1959. A much smaller enamel holloware factory was also assisted in the same way, though the result was less successful. The entry of cheap Japanese goods, in particular textiles and holloware, was restricted by import quotas as well as by the general duties. The ordinary consumer in all parts of Uganda, and not just in the areas supplied with electricity, has, therefore, had to pay considerably more for some of his basic necessities than he might have done without the dam.

Perhaps the major contribution which Owen Falls has made is one which was least considered in the 1940s and 1950s. The 'special industries' which figure so prominently among its consumers are unlikely to have been started if the dam had *not* been built and however negligable their contribution to the national product of Uganda may be, there can be no question but that they have been some help in diversifying and stabilising Uganda's exports. This is displayed by Table VI.

TABLE VI

UGANDA DOMESTIC EXPORTS — PRINCIPAL COMMODITIES — PERCENTAGE OF TOTAL VALUE

| | *Coffee* % | *Tea* % | *Cotton* % | *Copper* % | *Proportion of Total* % | *Total Exports Value* £m |
|---|---|---|---|---|---|---|
| 1954 | 33 | 2 | 52 | — | 87 | 40.6 |
| 1955 | 48 | 3 | 39 | — | 90 | 41.9 |
| 1956 | 39 | 2 | 48 | — | 89 | 40.4 |
| 1957 | 47 | 2 | 38 | 4 | 91 | 45.9 |
| 1958 | 46 | 2 | 40 | 5 | 93 | 45.4 |
| 1959 | 44 | 3 | 37 | 7 | 91 | 42.1 |
| 1960 | 41 | 4 | 36 | 9 | 90 | 41.6 |
| 1961 | 36 | 4 | 43 | 8 | 91 | 39.2 |
| 1962 | 54 | 5 | 22 | 10 | 91 | 37.6 |
| 1963 | 53 | 4 | 28 | 7 | 92 | 51.5 |

Source: *Uganda Statistical Abstract* 1959 and 1964

Table VI shows that copper has brought a welcome element of diversification into the heavy reliance on cotton and coffee exports whilst local textile and cement production ensure that however unfavourable the balance of payments may become in the process of development two items which would be bound otherwise to figure largely in the list of imports are taken care of because they are produced at home.

Finally, much of the spontaneous development of industry which ocurred in the 1950s has been the result of local *Asian* enterprise. It was they who had the necessary business experience even if they often lacked technical knowledge and only a handful had the gift of managing anything much larger than a wholesale business. Since independence, however, the government has followed a policy designed to enable Africans to participate in fields which had hitherto been the almost exclusive preserve of Asians. The result has been a crisis of confidence among Asians and despite government assurances they feel their future to be uncertain[13]. Africans have demonstrated their capacity for administration and for acquiring technical skills in a hurry, as witness the fact that the whole UEB now employs only 30 expatriates and that all but one of the senior administrative posts are held by Africans. But business acumen is not yet so well developed and if conditions no longer make it possible for Asians to play their customary role, then for the moment, the main hope for further development outside agriculture must be a government able to take active steps to promote it.

It may be asked 'was it on balance right or wrong to build the Dam' and our answer would be, first, that such questions are unprofitable since what is done cannot be undone. Secondly, if one wants to bring about economic development it is almost always better to do *something* rather than do nothing and the Owen Falls Dam at the time did not appear to be competing for funds with other projects. Nothing was left undone because the dam was built and the funds used for building it might never have been available for other purposes anyhow. The decision to build the dam was expected to induce others to take further decisions to create undertakings which would be made possible by a cheap supply of electricity[11]. The trouble has been that the decisions to create users for its output were not taken quickly enough. Even if one is sceptical on other grounds of theories expounding the need for balanced growth, it does seem that if a major electricity project in a small economy is not to be more of a burden than an asset, the decision to go ahead needs to be closely geared to specific proposals for the use to be made of its product.

[1] Parts of this paper are based on Miss Wilson's M.A. thesis accepted by London University in 1965 entitled 'Hydro-electricity in Uganda: a Geographical Study'. We wish to acknowledge gratefully help received from the Uganda Electricity Board. This chapter appeared originally in the *Journal of Development Studies,* Vol. 3, No. 4, July 1967.

[2] The project was, however, incorporated into the Egyptian scheme for long-term regulation of the River Nile by raising the dam one metre higher than was needed for hydro-electricity alone. The extra cost, nearly £1 million, was paid by Egypt.

[3] Kenneth Ingham, *The Making of Modern Uganda,* London, 1958.

[4] Winston S. Churchill, *My African Journey,* London, 1908, p.250.

[5] Report by Preece, Cardew and Ryder and Coode, Wilson, Mitchell and Vaughan Lee, consulting engineers, 1936.

[6] E. B. Worthington, *A Development Plan for Uganda,* Government Printer, Entebbe, 1947, p.xii.

[7] C. R. Westlake, *Uganda Electricity Survey,* Entebbe, 1947.

[8] C. R. Westlake, *op. cit.,* p.44.

[9] D. Walker, 'Power in Uganda: a review Article', *East African Economics Review,* Vol. 4, No. 2, January 1958, p.99.

[10] Communication by the Town Clerk.

[11] Cf. A. O. Hirschman, *Strategy of Economic Development.*

[12] Cf. C. R. Westlake, *Uganda Electricity Survey,* 1947, p.8. 'The present government of Uganda is acutely aware of the need for electrical development, and fully realises that the social and economic progress of the Protectorate demands the utilisation of the great potential wealth of the Victoria Nile'. See also G.B. Colonial Reports, *Uganda,* 1949, p.75. 'It will also play a most important part in raising the standard of living of the people; for it will make possible the establishment of industries which will increase the national income . . .'

[13] D. P. Ghai and Y. P. Ghai, 'Asians in East Africa; problems and prospects', *Journal of Modern African Studies,* Vol. 3, No. 1, 1965, pp. 35–51.

Chapter VII

# The High Dam at Aswan and the Politics of Control

D. C. WATT

The river Nile is the life blood of Egypt. Indeed to all intents and purposes it is Egypt. For eight hundred miles or more from Wadi Halfa to the southern approaches to Cairo its six-mile wide valley is virtually the only inhabitable land. Rainfall over Egypt is very slight. Without the Nile with its annual floods fertilising and bathing the land not even 15,000 of Egypt's 383,000 square miles would be able to support human life. The Nile's waters and the Nile floods depend on the movement of a belt of rain from the Equator northwards from March to July, the rain itself being most probably Atlantic in origin carried across Africa by the prevailing winds and precipitated over the Congo basin, the Ethiopian highlands and the high mountain masses of central Africa[1]. From these precipitations rise the two great tributaries of the Nile, the White Nile rising on a 6,000 foot plateau just north of Lake Tanganyika, and the Blue Nile with its principal tributary, the Atbara, rising in the eastern highlands of Ethiopia around Lake Tana. From the White Nile, which passes through the Great African lakes of Victoria and Albert and loses much of its volume in the swamps of the southern Sudan, the Nile derives its regular all-the-year round flow. From the Blue Nile and the Atbara come the floods.

Flood control and irrigation were the basis of the earliest Egyptian civilisation; and, as Egypt's population has advanced so the means of control have become steadily more sophisticated. Mohammed Ali, the Albanian founder of modern Egypt, first built a barrage in 1843 at the apex of the Nile Delta, south of Cairo, to raise the level of the Nile waters, so as to improve irrigation of land well above its normal level, though the barrage was not able to function properly until British engineers set it in order between 1882 and 1890. In the first years of the twentieth century British engineers began work on a dam at Aswan. Completed in 1905, its height was raised in 1912 and again in 1933 to hold back the excess of the annual floods for release in the slack seasons between the floods. This did not, however, take care of the potentially very wide variations in the total annual flow of the Nile. In 1874, 1879, 1892 and again in 1946 floods of up to 1·140 million cubic metres a day were recorded. In 1964 a flood came so high as to threaten the foundations of the High Dam itself. Other years have been low. Some decades have seen consistently high floods, others much lower. Between 1871 and 1908 and 1909–1945 the average annual discharge as measured at Aswan varied by 20 milliard cubic metres[2] or 25 per cent. From this the British hydrological engineers employed by the Egyptian government evolved the idea of long-term storage to even out the variations in flow not merely from one month or one season to the next but from one year, even one decade, to the next. The idea became known as 'century storage'.

The original idea of 'century storage' was to use Lakes Victoria and Albert on the White Nile and Tana on the Blue Nile as storage reservoirs, improving the flow of the White Nile by cutting a canal through the swamps of the southern Sudan where up to half of its flow is lost through seepage and evaporation, and constructing a series of dams along the whole length of the Nile both in the Sudan and in Egypt. As part of this the Owen Falls Dam was built a mile and a half below the exit of the Nile from Lake Victoria.

At this stage an Egyptian agronomist of Alexandrine/Greek parentage, M. Daninos, produced a proposal to construct a new and immense reservoir at Aswan to hold 186 milliard cubic yards of water, with a massive hydro-electric station and the capacity to bring nearly five million more acres under cultivation. Its great virtue, as he pointed out, would be that the whole project would be under Egyptian sovereignty, and could be embarked upon without lengthy and wearisome diplomatic negotiations[3]. This expectation turned out to be unusually sanguine.

The very fact of the untrammeled dependence of Egypt on the Nile led inevitably to speculation about the possibility of using the control of its floods at or near their source as a weapon of political control. This is not perhaps the place for a detailed account of the various occasions on which the idea had been mooted or become part of the coin of international diplomacy. One need only note that it was a French proposal, devised by the French engineer, M. Victor Prompte, to devise a storage dam on the

White Nile immediately below its juncture with the Sabat, which could be used to ruin Egypt 'by drought or untimely flood', which in 1893 set in move the events which were to lead to the Anglo-French crisis of Fashoda in 1898[4]; that concern over the future of the area around Lake Tana was the single most important motive in British policy towards Italy and Ethiopia from the Anglo-Franco-Italian agreement on the possible partition of Ethiopia in 1906, through the Italo-Ethiopian crisis of 1935 to the Anglo-Italian Agreements of April 1938[5]; that Count Ciano was threatening to turn the Nile against Britain in 1939; and that at the height of the Suez crisis in 1956 a Conservative M.P. in Britain, Major Patrick Wall, proposed that Britain should 'nationalise' the headwaters of the Nile against Egypt to bring her government to see reason[6].

For the Egyptians the unity of the Nile valley was not merely a slogan, but an ideal of the greatest importance to them. Under Mohammed Ali and his successors Egyptian rule had virtually extended as far as the Great Lakes. But the Mahdist rising in 1882 and the British suppression of the Mahdi State in 1898 had driven the limits of Egyptian sovereignty back to Wadi Halfa. Although the Sudan was in name under a joint Anglo-Egyptian condominium it was in fact a British colony. And the result of British rule was to preserve that sense of separateness from Egypt which had underlain the whole course of the *Mahdiya*. From 1898 to 1954 Sudan nationalism had grown under the benificent if sometimes puzzled gaze of the British-officered Sudan Civil Service. Separate Sudan military units had been gradually Sudanized, the University of Khartoum had produced a small anglicised élite, quite unlike the products of Cairo's al-Azhar University, and Sudanese political life and organization had grown up around the Sudani sects in a way for which the loose Sunni Mohammedanism of Cairo and Egypt offered no parallel whatever.

Still worse, the British encouraged the development of cotton cultivation through irrigation in the Gezira in direct competition with Egypt both for markets and for the use of the Nile waters. This was governed by an Agreement concluded in 1929 which limited the Sudan's water withdrawals to the surplus of the annual flood stored behind the Sennar dam. The natural flow was to be passed on to Egypt. The agreement was concluded after ten years of Egyptian allegation that the Nile was being robbed of its waters for the benefit of a British syndicate. In fact it gave the Sudan only 3 milliard cubic metres of water as against the very

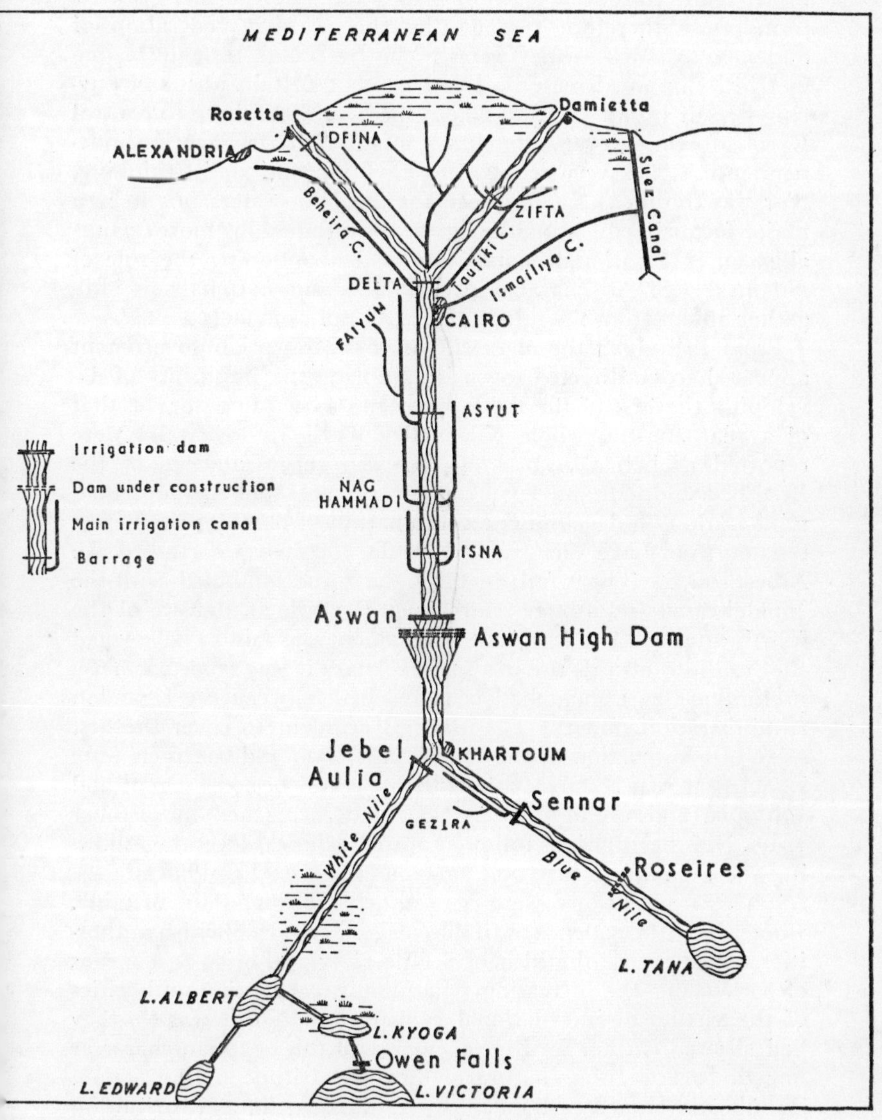


3. Diagrammatic representation of the Nile Conservation works
*Reprinted from Alan B. Mountjoy and Clifford Embleton,* Africa, A Geographical Study, *London, 1965.*

much more considerable volume of water allowed to Egypt. Its terms were therefore resented by the growing generation of Sudanese as much as they were by the Egyptians, though for diametrically opposed reasons. The idea that Britain was manoeuvring herself into a position where she would continue to control Egypt's fortunes, even though she might concede Egyptian independence, struck very deeply into Egyptian nationalist thinking. Nor was the idea that her need for the Nile waters might be a major factor in educating the Egyptians into looking more favourably on international cooperation and abandoning the role of *enfant terrible* so characteristic of emergent nationalisms altogether absent from the minds of British policy makers.

From 1949–1952 the main efforts of the Egyptian government appeared to be directed towards exploring the possibility of developing the first of the two approaches to century storage, that of a plan for the whole Nile valley. Early in 1949 talks were reported as being in progress between representatives of the British, Ethiopian and Egyptian governments and the Sudan administration on developing a combination of dams on the Ripon falls north of Lake Victoria, at Nimula, 100 metres north of Lake Albert, on the Tiscat falls in Ethiopia, to be combined with the Jonglei canal and a large reservoir on the Nile at Merowe by the fourth cataract. The Egyptian government was said to have voted £E4 million towards the cost[7]. In February it was reported that a preliminary agreement had been reached between the Ugandan administration and the Egyptian government to cover the first stage of construction of the Owen Falls Dam and contracts were issued that year[8]. Conversations then opened between Egypt and Ethiopia, and preliminary agreement was reached in October 1949[9]. New negotiations followed at the technical level the following year, ending in a second agreement in October 1950[10].

These negotiations were very much welcomed in Britain[11], where the little evidence available suggests that the British authorities hoped to use the idea of a Nile river authority as a means of convincing the increasingly nationalistic Egyptian authorities of the virtues of international cooperation. If this was so, they had allowed neither for the vehemence of the Egyptian character nor the overweening nature of Egypt's attitude to the Sudan. Trouble developed early in 1951 over the question of distribution of the Nile waters and Sudan's claims for a revision of the 1929 agreement. The Egyptians persisted in claiming total sovereignty over the Sudan, and only the actual experience of an independant Sudan could remove this obstacle. But long before this the

Egyptians had come to see in Britain the principal obstacle to their dream of the unity of the Nile valley. At the end of 1951 they denounced the Anglo-Egyptian Treaty of 1936 and bitterly attacked the Anglo-Ethiopian Treaty of 1899 which gave Britain a veto on projects to dam Lake Tana. With this it could be said that the first attempt to use the Nile waters question to 'control' Egypt can be said to have failed.

With the Army coup d'état in July 1952 in Egypt and the advent of the Nasser-Neguib régime, this impression was confirmed. The notion of using the African Great Lakes for storage seemed to the newcomers to be dependent on the successful coordination of too many interests. Moreover it left the British with too big a say in matters, and too many opportunities to control the headwaters or otherwise interfere with the water on which Egypt depended. The new régime looked for autarchy in matters economic, so far as this could be achieved.

To the incoming régime the Daninos scheme for a High Dam not only seemed to solve the century storage problem on autarchic lines. It also promised hydro-electric power for industrialisation. The immense urgency they felt to do something to tackle the modernisation of Egypt and the problem of industrialisation, over-population, etc. was decisive. Within two months of the coup, the new régime had adopted the High Dam scheme for their own.

At this stage the exigencies of inter-Arab politics presented them with a sudden windfall. As part of its campaign to achieve a rehabilitation of Germany with world opinion, and inspired also by the need to establish itself as the sole legitimate government of Germany, the West German government had concluded in 1952 a restitution agreement with the state of Israel. The reaction of the Arab League to this had been to threaten a boycott of West German goods throughout the Middle East. To avert this the Bonn government sent the Federal Under-Secretary of Economic Affairs, Herr Dr. Westrich, to Cairo in February 1953. His visit was followed by the announcement that Mohrad Fahmi, the Egyptian Minister of Public Works had invited a commission of experts from West Germany to examine the technical feasibility of the High Dam project. A German mission headed by the director of the association of barrages of the Ruhr, Dr. Preuss, and an engineering expert from the two construction firms, Hochtief of Essen and Union Brüchenbau of Dortmund, duly arrived in Cairo and after a few months work not only reported that the High Dam was technically feasible, but also drew up blueprints for its construction at an estimated cost of £E100 million with a reservoir capacity of 125,000 cubic metres and a very large hydro-electric capacity.

The submission of a scheme on this scale inevitably raised the question of finance. Only one obvious source presented itself, the World Bank, to which the Egyptians duly applied for a loan of £E60 million. There followed a prolonged examination of the project, both from the technical and from the financial point of view, by experts appointed by the World Bank. By the end of August 1955 the World Bank authorities declared themselves satisfied that the project was 'technically and economically sound'. In the meantime the Egyptians had been negotiating with the British Treasury for an accelerated release of the £150 million sterling owed to the Egyptians at the end of the Second World War held in a blocked account in the Bank of England. The Egyptians wanted the rate of release upped from £10 million sterling to £25 million a year over the six year period while the Dam was being constructed. Preliminary agreement was followed by the dispatch of an Egyptian mission to Britain to discuss British engineering participation in the construction. In September 1953 an Anglo-German consortium was formed, and in October, Sir Alexander Gibb, the well known British construction engineer was appointed consulting expert. On October 4, 1955 the Egyptian Government formally applied for a loan from the World Bank.

In the meantime, however, the problem of relations with the Sudan had risen once again to bedevil things. In October 1952 the western press had reported the conclusion of an agreement between the two countries, only to have the reports denied in Cairo[12]. Serious negotiation does not seem to have begun until the spring of 1954 when a new agreement[13] was reported to have been reached in Cairo. Sudan's anxiety about the High Dam, however, necessitated new talks later in the year as the Sudanese were unwilling to pursue their scheme for a dam at Roseires until they knew what Egypt wanted for the High Dam. Although Egyptian consent was obtained for an increase in Sudanese water drawing from the Jebel Antra dam, the general question of distribution remained unsettled. In April 1955 the talks were suspended in an atmosphere of mutual recrimination. The basic difficulties were twofold. On the question of the division of the 84 milliards cubic yards of Nile water annually passing Aswan the Sudan claimed 35 milliards and the Egyptians would only grant 8 milliards. The Egyptian attitude was very uncompromising, and they refused to concede either of the two principles advanced by the Sudan as a base for distribution, relative population or relative acreage under cultivation. Instead they alleged that the Sudan minister for irrigation had been printing and distributing anti-Egyptian propaganda leaflets while in Cairo. Fresh talks followed in May 1955 between the Sudanese premier el-Azhari and President Nasser which succeeded in getting the negotiations going again. But no new Egyptian proposals emerged from these new talks, not even when an Egyptian delegation reached Khartoum. The dispute with the Sudan remained, to complicate and bedevil the next stage of Egyptian involvement over the High Dam.

The Egyptian application to the World Bank raised immediately the question of how that institution was itself going to raise funds on a scale sufficient to cover Egypt's needs. In practice they could only be raised in the United States and Britain, the two major capital markets of the non-Soviet world, with the support of the American and British governments. The Egyptian application to the World Bank presented these two governments at once with a dilemma and, so it seemed, an opportunity. Since 1950, American and British policy in the Middle East, though differing sometimes on methods, had agreed on pursuing two aims: the maintenance of a balance of armaments between Israel and her Arab neighbours, and the exclusion of the Soviet Union by the bringing together of those Middle Eastern states which had a common border with the Soviets into mutual security agreements, the so-called 'Northern Tier'. These aims had suffered a severe setback in September 1955 with the conclusion of a major arms deal on credit between Egypt and Czechoslovakia, in which the latter state was clearly acting as agent for the Soviet Union, an arms agreement which, at the time of its signature, gave the Egyptians technologically advanced weapons, especially in the air, more advanced that any either Britain or France, let alone Israel, maintained in the Middle East. The United States failed to secure any revision in the agreement despite the dispatch of a special mission to Egypt, under Mr. George V. Allen of the State Department, to see the Egyptian authorities, and a sharp exchange of words between the U.S. Secretary of State, Mr. Dulles and his Soviet opposite, Mr. Molotov, at Geneva, and of notes between President Eisenhower and Mr. Bulganin in October[14].

For the British government the growth of Egyptian influence and prestige effected by the Soviet-Egyptian arms deal presented additional problems. In the Lebanon, President Chamoun, was under strong pressure to move his western-oriented government into the Syria-Egyptian defence agreement. They were reaching a critical phase in the attempt to persuade Jordan to join the powers of the 'Northern Tier' in the Baghdad Pact. Cairo radio's propaganda and Egyptian subversion were causing them anxiety in Libya, with which there was a British defence agreement and to whose territory the armoured component of the Middle Eastern strategic reserve had been evacuated with the conclusion of the Anglo-Egyptian agreement on the Suez Canal base, to the vitally important oil sheikdoms of the Gulf. The Iraqi government under Nuri es Said were constantly protesting against the campaign of propaganda and abuse levelled by Cairo radio against Iraqi membership of the Baghdad pact. And Britain still felt obligated towards the Sudan. Finally the British premier, Anthony Eden, had staked a good deal of personal prestige on the conclusion of the Anglo-Egyptian agreement on the evacuation of the Suez Canal, and was already facing severe attacks from an organized section of his party's right-wing.

A further complication was that the Soviet ambassador in Cairo, M. Daniel Solod, was already being quoted in the Cairo press (the Soviet press, by and large, were discreetly silent), as having offered Soviet financial aid towards the construction of the High Dam. At this early date this was clearly more an Egyptian device to extract the maximum concessions from the West rather than a serious proposition[15]. The reports were in fact accompanied by outright threats that, unless the United States used its good offices to persuade the World Bank to grant the Egyptian request, Egypt would feel obliged to accept the Soviet offer.

This campaign was maintained in the Egyptian press throughout November 1955, while the Egyptian Finance Minister, el Kaissouni, was conducting talks in London and New York on the terms on which the World Bank and the British and United States governments would be prepared to meet the financial needs of the High Dam. On November 20, M. Solod held talks with President Nasser and with Wing Commander Saleh Salem, the minister in charge of the Dam project. The Polish government was also reported to have made an offer of financial and technical assistance[16].

At the same time Anglo-American discussions on the Egyptian application were taking place in London and Washington[17]. To avoid the further spread of Soviet influence in Africa, and to control Egypt's associations with the Soviet bloc, it was decided to support the Egyptian application for a loan on three conditions. Egypt was to concentrate her internal resources on the project. She was to prevent further inflation of her currency. She was to reach agreement with the Sudan on the division of the Nile waters. It emerged later that one further condition had been agreed on. No aid was to be sought for the Dam from Soviet Russian or Soviet bloc sources[18]. On this understanding final talks were held between Herbert Hoover Junior for the State Department, Sir Robert Makins, Mr. Eugene Black and el Kaissouni in Washington. On December 2, World Bank agreement to the Egyptian request was announced. The World Bank was to loan the Egyptian government $20 million to defray the foreign exchange costs of the first stage of the Dam. In addition the United States and British governments would make outright grants of $56m and $14m respectively 'subject to legislative authority'[19]. 'Sympathetic attention' was promised to the granting of a further loan of $130m to cover the second stage of the Dam.

The implications of this offer were very clear. So long as the Aswan Dam was under construction, the Egyptian economy was to be so organized as to make its financing the overriding priority. This had considerable implications both for Egyptian plans for a more rapid industrialisation and for her commitments to pay for her Soviet arms purchases by cotton deliveries. At a cost of $20m per annum, (President Eisenhower's estimate)[20], the United States was proposing to purchase a means of continuing her otherwise slipping control of Egyptian foreign policy. As a proposal it was open to one, essentially insuperable, objection. While its motives were only too plainly understood in Cairo, they were impossible to explain publicly to Colonel Nasser's enemies, either in Britain and the United States or in the Middle East. Over the next six months Egyptian diplomacy was directed on the one hand to the attempt to obtain a modification of the conditions attached to the original offer, on the other to demonstrating ostentatiously Egyptian independence and determination to follow her previous policy. At the same time, the granting of the loan and the entire dam project was the subject of heavy attack both in the United States Congress and the British parliament, and on the part of Iraq, Turkey and the Sudan.

The initial reactions of the Egyptian press at once reflected the unwillingness of the Egyptian government to accept the terms of the World Bank and of the American and British governments. The loans were described variously as an 'offer' open to Egypt to accept or reject, or as the 'first stage' to which Egypt's reply would be 'the second step'[21]. On the first day of the new year, President Nasser was reported as having discussed the matter with the British ambassador who was then about to leave for London[22]. The World Bank was asked to clarify the condition they had made that all contracts should be let on a competitive basis. The occasion was taken solemnly to announce that Egypt could accept no limitation on her sovereignty[23]. Mr. Black then flew to Cairo for talks with el Kaissouny. On February 10 it was announced that they had reached agreement on the level of interest to be paid, and that Egypt had withdrawn her objections to comparative tendering. But the Egyptians also announced that it would be two to three years before they would start drawing on the loan[24]. Diplomatic effort was also exerted to avoid the implication that Anglo-American agreement on financing the second stage of the loan would be dependent on evidence of Egyptian good behaviour[25]. The legislative pressures of both countries, however, made such an implication inevitable. It was underlined by the agreement reached between Sir Anthony Eden and President Eisenhower on the occasion of the former's visit to Washington in February 1956, when it was agreed that the outcome of the current talks between President Nasser and Mr. Black should be taken as evidence of Egypt's worthiness 'to cooperate with us'. If President Nasser's attitude was that 'he would not cooperate, we would both have to reconsider our policy towards him'[26].

The Egyptian government in fact showed itself extremely disinclined to provide evidence of good behaviour of the kind required by the British government. Between February and June 1956, the British position in the Middle East deteriorated markedly. Highlights in this deterioration were the stoning to which the official car of Mr. Selwyn Lloyd, the Foreign Secretary, was subjected in Bahrein, and the dismissal of General Glubb from his post as commander of Jordan's Arab legion; but they were only symptomatic of a much more far reaching deterioration in the British position. Eden for a time expressed himself as deeply worried by the possible spread of Soviet influence in the Middle East, and addressed himself very firmly on the subject to Marshall Bulganin and Mr. Khrushchev on the occasion of their official visit to London in April 1956[27]. But the course of events in the Middle East, especially in Jordan and Syria, showed that there was in fact little need for the Russians to play an actively disruptive role in the Middle East, when the Egyptian government were doing so well on their own.

This deterioration in the British position was taken by Eden's critics in his own party as justifying all the criticisms they had ever made of his efforts to reach an accommodation with Egypt. Beginning in February 1956 with an attack on Egypt's failure to reach agreement with the Sudan, Eden's critics launched both in parliament and in the correspondence columns of *The Times,* a continuous campaign against the whole concept of the High Dam. In April 1956 they seized astutely on the hydrological issue. It was alleged that the evaporation rate at Aswan would be so high that far more water would be lost than the Dam could possibly conserve. The critics, with similar astuteness, returned over and over again to the Foreign Office's own past proposals for an international Nile authority, varying and justifying this by accusations that not only the interests of the Sudan but also those of Britain's colonies in equatorial Africa were being allowed to go by default[28]. The campaign forced the spokesman of the Foreign Office to say that the High Dam was not inconsistent with the establishment of an international authority for the Nile valley and that the British government believed that one might well come into being[29]. An official Egyptian protest that the Commons had no right to discuss a question which was solely the responsibility of those who lived in the Nile valley[30] was ignored as was the evidence of Mr. Hurst, despite his generally acknowledged status as the leading hydrological expert on the Nile waters and originator of the whole idea of century storage. The general picture created was one of a British government, pandering in a craven manner, to the Pharaonic ambitions of an Anglophobe dictator. By June 1956 the High Dam had few public defenders in Britain outside the ranks of the government and Whitehall.

In the middle of June, Mr. Selwyn Lloyd approached the State Department through the British embassy in Washington, on the future of the Aswan Dam project[31]. The inquiry showed that the American authorities had grave doubts about the seriousness of the Egyptian application. The information appears to have crystalised British government opinion against the project. By the middle of July, according to Lord Avon's version, the Cabinet had come to the conclusion that the whole Aswan Dam project, loan and grant together, was likely to be 'increasingly onerous in finance and unsatisfactory in practice'[32]. How to be released from the obligation was not decided. Eden, so he says, would have preferred to 'play it long', that is, to allow the issue to perish from neglect and inanition[33]. Such was not however Mr. Dulles' way of doing things.

In America in the meantime the dam was being publicly subjected to a campaign of political criticism and denigration quite as strong as that launched against it in Britain by the Suez group. The substance and inspiration of the campaign in America were however different. The core of the opposition came, in the Senate

Appropriations Committee, from southern senators concerned over alleged Egyptian export subsidies for cotton, in competition with American cotton, which was currently selling at a rather low price on the world market, and from the pro-Israeli elements opposed to anything which might strengthen Egypt[34]. By the middle of June 1956 the Eisenhower administration had become frankly dubious as to its chances of obtaining senatorial approval for the American grant in aid except with an all-out effort, which they had good domestic reasons for not wishing to indulge in an election year[35]. The matter was complicated by the fact that the delay of six months between January and June 1956 had carried them into a new financial year. Had their offer been accepted before June 15, it could have been financed under moneys already appropriated for the preceding financial year. But under the provisions of the constitution those moneys lapsed on June 15, and fresh application had to be made to Congress. On July 6, the State Department announced that the moneys appropriated for aid to Egypt in the financial year 1955–56 had been directed to other purposes; and on July 15, the report of the Senate Appropriation Committee on the new Mutual Security Act included a clause prohibiting the use of funds for the Aswan Dam without their approval. Such a stipulation was in fact unconstitutional and President Eisenhower refused to accept it. But it only underlined the conclusion which he and his administration had already reached[36].

This conclusion was reinforced by two other sets of developments. The members of the Baghdad Pact, especially Turkey[37] and Iraq had begun as early as January 1956 to protest at the favourable treatment, for so they regarded it, given to Egypt despite its poor record of cooperation with the West. The Sudan government too had threatened to veto the whole project unless its rights were fully safeguarded. It was alleged that the construction of the High Dam would rob the Sudan of a hydro-electric potential of some 550,000 h.p. by submerging the second cataract of the Nile and reducing the drop at Zemna. On March 2, the Sudan authorities published a communiqué demanding the protection of the interests of the Sudanese inhabitants of Wadi Haifa, doomed to be totally submerged behind the Dam. The Sudan's cause was, as we have seen, vigorously taken up in the British House of Commons by right-wing critics of British policy.

It is at this stage that the Egyptian motives themselves become difficult to assess. The Egyptian government began, as we have seen, by applying to the World Bank for a loan, and using the bogey of Soviet aid as a lever to force more concessions out of the Bank and its Anglo-American backers. They continued that policy in the first few months of 1956, at least up to the beginning of April. By then, when Colonel Nasser was reported as saying that he was holding the Soviet offer of assistance in financing the Dam in case there was a breakdown in negotiations with the West, there were signs that he was beginning to resent the pressure to which he was being subjected[38]. From the British side it appeared as if he was making even more and more difficulties[39]. And the Americans had concluded by June 1956 that Nasser was holding onto the western option only as a means of trying to get a better offer from the Soviet Union[40]. In that month, Mr. Shepilov, the Soviet foreign minister, visited Cairo for the celebrations which accompanied the withdrawal of the last British troops from the Suez Canal base. He spoke largely of Soviet loans of up to $400m. for development purposes. And on June 20, Colonel Nasser saw Mr. Black in Cairo. His counterproposals then were so largely unacceptable and so extraordinary as to lead the Americans 'to conclude that he was not seriously interested'[41].

By this date he had obviously begun to plan his new coup against the Suez Canal company, with whom up to the end of April 1956 he had been in amicable negotiations. His behaviour in the crucial week before the nationalisation of the Canal suggests that he had genuinely believed in the Russian offer of an alternative loan. But their immediate denials that a formal offer to finance the claim had in fact been made show that if he did believe in it, he was to be very quickly disappointed. The final approach to the Americans on July 19 was so hurried that one does not have to accept Mr. Dulles' story of intercepted telephone calls[42], to conclude that the Egyptian government already expected their approach to be turned down; though they cannot have expected that Mr. Dulles would play so much into their hands by so brusque and public a rejection, couched in terms which could so easily be, and were, represented as a direct attack on Egypt's national pride. To Mr. Dulles the opportunity to 'cut Nasser down to size' seemed too good to be missed. The unilateral American action left the British no option but to follow suit. And the way was clear for the nationalisation of the Suez Canal[43].

The American decision was taken very suddenly. On July 13 Dulles had warned the Egyptian embassy that America was not in a position to decide on the Aswan loan, Congress's views being unpredictable, and the administration's views on the merits of the matter having changed[44]. The Egyptian ambassador in Washington, then in Cairo, Ahmed Husein, appears to have persuaded Colonel Nasser to make one last appeal for western aid. On July 17 he left Cairo for Washington announcing that he was returning to sign the Aswan loan agreement. A meeting with Mr. Dulles was arranged on July 19. It was not until that morning that the American Secretary of State finally decided to tell Husein that the loan offer was to be withdrawn. The British ambassador was only informed an hour before the meeting. At the interview itself, Dulles reportedly said that the United States would not be blackmailed by Egypt's threats to turn to the Soviet Union[45].

Later evidence suggests very strongly that it was the failure of his attempt to use the Aswan loan as a means of controlling Colonel Nasser, and the latter's attempt to turn the instrument of control against him, so as to control America, the donor, rather than submit to American control, which most incensed Mr. Dulles against Colonel Nasser. Dulles was convinced the Russian 'offer' was unreal, and determined to show it up as such. Nasser, in his view, was playing both sides. To discomfort him would not only be a salutory reminder to him where real power lay. It also would stand as a lesson to any other state which tried to play both sides against the middle to its own advantage[46].

With the outbreak of the Suez crisis, the issue of the Aswan dam faded for the moment into obscurity. The Russians were unwilling to finance the dam for the time being[47] and tentative soundings in Japan, Switzerland, Italy and West Germany produced the occasional headline but nothing more. It was not until Egypt had united itself with Syria in the U.A.R., survived the crisis of July–August 1958 which threatened at one time to lead to a direct United States clash with the Soviet Union, and Anglo-American forces evacuated the Lebanon and Jordan, that negotiations on the High Dam reopened between Egypt and the Soviet Union. This time the Soviet authorities, who were turning their main political interests again towards Europe and the reopening of the Berlin crisis appear to have treated the issue on purely economic terms. It is true that they insisted on the withdrawal of the last elements of West German participation. But from then on the negotiations were at the technical level between Soviet and Egyptian engineers. Nor have the Soviets made any real propaganda effort to exploit their role in the construction of the High Dam, being apparently content to let President Nasser get what mileage out of it he could.

With the conclusion of the Soviet-Egyptian agreement and the exchange of contracts in September 1959, the political side of the history of the High Dam at Aswan might have been expected to be over. But this would be to ignore the problem of Egyptian relations with the Sudan and the idea of a single authority for the Nile valley. As has been seen earlier this was for long advocated by the British in such a way as, to put it at the very least, to leave the Egyptians with the impression that their main motive was to use it as a means of controlling Egypt and extracting good behaviour from her. But the 'unity of the Nile valley' had had a long life as an Egyptian political concept before the British came to see political advantage in interpreting it ideologically. In Egyptian parlance it meant a re-establishment of that Egyptian sovereignty over the Sudan which had been destroyed by the Mahdist rising of 1885. Despite all evidence to the contrary the Egyptians had persisted in closing their eyes to the long history of Sudanese antipathy to rule by Cairo, and had persuaded themselves that Sudanese nationalism was the invention of British imperialists and their pawns which would soon collapse once the protecting power was withdrawn. They had thus treated the Sudan's claim on the Nile's waters as something not to be taken seriously; and were therefore unpleasantly surprised in July 1958 when following a new breakdown in talks with the Sudanese authorities the latter began to divert Nile waters to the Gezira from the Sennar dam without waiting for Egyptian agreement.

The Sudan's action was intended to underline the three conditions to which they insisted on obtaining Egypt's agreement before themselves agreeing to the construction of the High Dam. Firstly, the Sudan's share of the Nile flow as calculated at Aswan had to be determined before construction began. Second, Egypt must formally recognise the Sudan's right to take water from the Nile and its tributaries and not pretend as heretofore that Egypt was merely condoning Sudanese action as an act of grace. Third, the cost of resettling the Sudanese inhabitants of Wadi Halfa must be borne by the Egyptian exchequer. The Egyptians replied with a violent exchange of notes followed by a full scale campaign against the Sudanese government for pandering to the U.S. State Department's claim for a Nile valley authority, and aligning themselves with the 'colonial feudatories of Uganda and Tanganyika'[48]. The presence of a World Bank mission in Khartoum was linked with this. 'All the weapons of cold war and propaganda which the Baghdad Pact used against Egypt have been moved to Khartoum, including British newspaper men known for their anti-Egyptian attitude . . . the Sudan government is the victim of a conspiracy', wrote *al Ahram*[49].

The Egyptians were rather less than candid in their attitude to the Sudan. The average annual flow at Aswan was estimated at approximately 84 milliard cubic yards, of which under the agreement of 1929 Egypt used 48 milliards and the Sudan 4½. In the negotiations which broke down in January 1958, the Egyptians had begun by offering the Sudan a mere 6 milliards, raising the figure only under great pressure to 13 milliards. The Sudan now planned to revive Mr. Hurst's original 'century storage' plan, or at least those parts which applied to the Sudan. In this way they hoped to add some 28 milliards to the flow at Aswan. 9½ milliards were to be gained by stopping leakages, the remainder by the construction of new reservoirs and the addition to existing ones. Dams were to be built at Lakes Albert, Kiga and Victoria. A canal, the Jonglei scheme, through the Sudd wastes would add 13 milliards a year hitherto lost by evaporation and seepage. A dam in Ethiopia at Lake Tana would be coupled with a tunnel connecting Lake Tana with the headwaters of the river Balas. New dams were to be built at Sabat, Roseires on the Blue Nile, or the Atbara at Khasm el Girba, at Sabaloka, Merowe and Semna. With all this the Sudan reckoned that although 12 milliards would be lost by evaporation in the new reservoirs, and 1½ milliards to East African irrigation, they could store 80 milliards, and it would not be necessary to construct the High Dam at all; though there would have to be some increased capacity at Aswan.

This was turning the tables on the Egyptians with a vengeance: the lesson was reinforced by the overthrow in September 1956 of the Sudan government by a military junta led by General Abbud, which remained resolutely impervious to Egyptian appeals for Arab unity. The Sudan's claims, however, were themselves too far reaching for any international organization to finance while Egypt remained an independent state. And the conclusion of the Soviet-Egyptian aid agreement in January 1959 made it clear that Egypt was determined and able to go ahead with the construction of the High Dam. New negotiations opened in October 1959 in an atmosphere of greatly increased realism on both sides, reinforced by the World Bank's refusal to finance the construction of a dam at Roseires until the Sudan had reached agreement with Egypt on the disposition of the Nile waters.

Once again Ethiopia and Britain intervened. The official *Ethiopian Herald* complained on June 20 1959 that Ethiopia supplied '84 per cent of the Nile's waters and most of the alluvial soil', and called for 'an international approach conforming to and guided by international practice in all discussions of projects designed to utilise the Nile waters'[50]. In a Note of August 11, acting on behalf of the governments of Kenya, Uganda and Tanganyika, the British government addressed the Sudan, Egypt, Ethiopia and the Congo on the irrigation needs of her east African colonies, now well on the way to independence. On October 8, the Foreign Office issued a statement, subsequently denied, claiming yet again that 'a comprehensive settlement of the Nile water's question will require a conference of all riparian states'[51]. The Cairo press were stirred to fresh denunciations of Britain's attempts to intervene in a matter which does not concern her[52].

The intervention, however, was not without its effect in reminding both the Sudan and Egypt of the dangers of allowing matters to be drawn out any longer. Talks opened in Cairo on October 10, in an atmosphere of 'almost suffocating goodwill' as *The Times* correspondent sourly remarked[53]. Final agreement was reached and signed on November 8, 1959.

The terms of the agreement showed that the Egyptians had been forced greatly to moderate their claims. The Sudan was to get 18½ milliard cubic yards of water, loaning 1½ milliard of this annually until 1977 when it was estimated her own irrigation works would be completed, and the completion of the High Dam would enable Egypt to repay the water borrowed over a further ten years. The cost of any development of the upper Nile was to be shared equally and the extra waters saved to be equally divided between the two countries. A permanent joint technical committee was to negotiate with the other riparian states, and the two countries agreed to maintain a unified front to their demands for a share of their waters. Egypt was to pay £E15 million as compensation to those Sudanese dispossessed at Wadi Haifa (an increase of £E1 million on the original Egyptian offer).

Technical talks have in fact been in progress for several years with the other riparian states. But until the High Dam is completed there will be little money and no water over for any further development of the Nile such as the Jonglei canal scheme. Ethiopia has remained outside the scope of these talks, her own plans still remaining at the same level as that of 1952, projects without plans, funds or energy to realise them. As and when Ethiopia abandons its position of inactivity serious talks can be expected to begin. For the Nile waters are a weapon of the weak who need them only a little against the strong, Egypt, whose whole life depends on them.

The course of the negotiations on the financing and constructing of the High Dam at Aswan provide an instructive example of the political implications of water control on international rivers. There can be little doubt that among the motives which led the British government to plead for the establishment of an international authority for the Nile waters on which all the riparian states should be represented, there was included the notion that Egyptian nationalism could be domesticated, and instructed in the complexities of contemporary international politics; that its nativist expansionist national-centric aspects could be 'controlled', and that its energies could be diverted into that kind of 'constructive' administration, which the paternalist-colonialist philosophy of latter-day British imperial sentiment used to contrast with the 'destructive' nationalism of nationalist politicians. There can be equally little doubt that among the virtues of the High Dam scheme in the eyes of the young officers who seized power in Egypt in 1952, one of the most important was that the whole scheme could be kept under Egyptian sovereignty and control. When the necessity of securing external finance for the High Dam revived the possibility of establishing some kind of 'control' over Colonel Nasser's regime, it was this issue again which inspired both the original American offer of aid, and its later withdrawal. Equally it was Colonel Nasser's refusal to accept such control that precipitated that withdrawal.

Yet it is interesting to observe that on their side the Egyptians were unable to make their own position on the Nile pre-emptive, despite their attempts to do so. In the negotiations with the Sudan they were forced to concede virtually all that the Sudan government demanded, yielding to the advantages which geography gave to the Sudan what they were not prepared to concede to Britain. Ethiopia by contrast has so far proved totally unable to use her position of geographic advantage as source of the main Nile floods. Partly this may be attributed to lack of will-power; but that will-power is itself dependent on technical feasibility, and until there is established at Lake Tana or on the course of the Blue Nile up to the Sudan frontier a complex of dams which make control of the Nile flow possible, there is little the Ethiopian government can do, even diplomatically.

For it should be noted that the question of control of the Nile waters has always been, save for the brief episode of Fashoda, entirely a question of diplomatic bargaining. The actual use of such control against Egypt has only been mooted, never practised. Despite the views of the Suez Group it is inconceivable that public opinion in Britain would have supported any attempt to use the control of the Nile waters as an instrument of actual economic pressure against Egypt. A survey of the Nile published in 1957 by the Sudan government suggests that such control could only have been exerted in the Sudan. The use of the dam at the outlet to Lake Victoria would only restrict about 10·5 milliard cubic metres out of an average annual flow of 85 milliard cubic metres past Aswan. About 72 milliard of the total average annual flow originated in fact in Ethiopia[54]. So unreal in fact were the fears of the Egyptians and the threats of Major Wall, M.P., *et hoc genus omnes.*

The issue of control of the Nile waters and of the High Dam turned in fact on the legitimisation of the demands of the Egyptian government for control over their most vital national asset. The final outcome was more a vindication of their stand for independence than that legitimisation of their claims for which they strove. They found themselves forced to concede the claims of the Sudan, even to take part in a systematisation of contacts with the other riparian states, though at a much lower level than originally proposed by the British. In return, however, they secured the High Dam and a freedom from a more general international control, plausibly denounced by Colonel Nasser as 'collective colonialism'. The slow filling of the lake behind the High Dam is their reward.

[1] W. E. Hurst, *The Nile,* London 1952, pp. 255–58.

[2] *Ibid.,* p.296.

[3] This summary is based on Tom Little, *High Dam at Aswan,* London 1965, pp. 28–35.

[4] See especially E. N. Sanderson, *England, Europe and the Upper Nile, 1882–89,* Edinburgh 1965, pp. 142–144; W. L. Langer, *The Diplomacy of Imperialism,* 2nd Edition, New York 1951, pp. 127, 135, 558–560, 575.

[5] See Carlo Giglio, 'La Questione di Lago Tano', *Rivista di Studi Politici Internazionali,* Anno XVIII (1951), No. 4., pp. 643–686; D. C. Watt, 'Gli accordi mediterranei anglo-italiani del Aprile 1938', *idem,* Anno XXVI (1959), pp. 51–76.

[6] Hansard's Parliamentary Debates (Fifth series), House of Commons, vol. 558 at col. 52, 12 September, 1956.

[7] *N.Y. Times,* 24 January 1949; *Bourse Egyptienne,* 8 February 1949; *Egyptian Gazette,* 9 February 1949.

[8] *Financial Times,* 11 February 1949; *Egyptian Gazette,* 20 May 1949.

[9] *Bourse Egyptienne,* 11 August, 18 October 1949.

[10] *Ibid.,* 27 February, 3 October 1950.

[11] Mr. Hector MacNeill, Minister of State in the Foreign Office called it 'A magnificent exercise in international cooperation', *Financial Times,* 11 February 1949.

[12] *The Times,* 14 October 1952.

[13] *Ibid.,* 14 April 1954.

[14] Dwight D. Eisenhower, *The White House Years,* vol. II, *Waging Peace 1956–1961,* New York 1965, pp. 25–26.

[15] The first reports of M. Solod's alleged offer to give Egypt 'all the help they wish' appeared in the Egyptian press on October 18, 1955, citing an interview given by the Egyptian ambassador in Washington to selected American and Arab journalists, *Bourse Egyptienne,* 18 October 1955. It was accompanied by attacks on the United States by Nasser's principal propagandists for trying to 'enfeudalise Egypt in the western defence bloc'. See Anwar Sadat in *Al Tahrir,* also *Al Gumhuriya,* 18 October 1955 cited *Bourse Egyptienne, Financial Times,* 19 October 1955.

[16] *Bourse Egyptienne,* 22 November 1955.

[17] Anthony Eden, *Memoirs,* vol. III, *Full Circle,* London 1960, p.420.

[18] *Ibid., op. cit.,* p.420; Eisenhower, pp. 30–31.

[19] The text of the American offer is printed in *Department of State Bulletin,* vol. 33, 26 December 1955, pp. 1050–51.

[20] *Op. cit.,* p.31, footnote 10.

[21] *Bourse Egyptienne,* 26 December 1955; an Egyptian government spokesman cited in *Financial Times,* 28 December 1955.

[22] *Financial Times,* 2 January 1956.

[23] *N.Y. Times,* 3 January, 1956.

[24] *Financial Times,* 15 February 1956.

[25] *Times,* 25 January 1956.

[26] Eden, *op. cit.,* p.335.

[27] Eden, *op. cit.,* pp. 357–359.

[28] See the interventions of Lord Vansittart, 29 February, and Lord Killearn, 14 March 1956 in the House of Lords, *Parl. Deb.*, House of Lords, vol. 196, cols. 55–56 and cols. 387–88; the letters to the *Times* of Hugh Frazer, M.P., Professor Jack, Mr. M. W. N. Allan, and Mr. John Paget, *The Times,* 16 April, 18 April, 21 April, 1 May, 4 June 1956; the interventions of Hugh Fraser and Captain Waterhouse, 21 March, 25 April in the Commons, *Parl. Deb.*, House of Commons., vol. 550, col. 1241, vol. 551, cols. 1755–57, and of Hugh Fraser, Richard Stokes, Julian Amery and Bernard Braine, 18 May 1956, *Parl. Deb.*, House of Commons, vol. 552, cols. 2375–2411.

[29] *Ibid.*

[30] See *Bourse Egyptienne,* 22 June 1956.

[31] Eden, p.421.

[32] *Ibid.*

[33] Eden, p.422.

[34] The type of arguments used may be seen in U.S. Senate, Committee of Appropriations, 84th Congress, second session, *Financing of Aswan Dam in Egypt, Hearings,* Washington, 1956.

[35] Eisenhower, p.31.

[36] *Ibid.*, p.32, footnote 12; see also Dulles to Eisenhower, 15 September 1956, cited *ibid.*, p.33.

[37] See, for example, the comments of the famous Turkish journalist, Yalman in *Vaten,* cited in *Journal d'Orient,* 4 January 1956. For Iraqi complaints on similar lines, see Eden, p.421.

[38] See, for example, his complaint to the *N.Y. Times,* 2 April 1956, that Britain was stirring up the Sudan to delay work on the High Dam.

[39] Eden, p.420.

[40] Eisenhower, p.31.

[41] *Ibid.*, p.32.

[42] Dulles to Eisenhower, 15 September 1956, printed in *ibid.*, p.33.

[43] See James E. Dougherty, 'The Aswan Decision in Perspective', *Political Science Quarterly,* vol. LXXIV, (1959), No. 1, pp. 21–45.

[44] Eisenhower, p.52.

[45] Robert Murphy, *Diplomat among Warriors,* London 1964, pp. 459–60.

[46] Dulles Press Conference, 2 April 1957, *N.Y. Times,* 3 April 1957.

[47] The Soviet Embassy in Cairo spent an embarrassed week explaining that the Soviet offer of aid had not actively gone so far as to cover the financing of the High Dam.

[48] Cited in *Bourse Egyptienne,* 1 October 1958.

[49] Cited in *Egyptian Mail,* 11 October 1958.

[50] Cited in *The Times,* 11 June 1959.

[51] Cited in *N.Y. Herald Tribune* (European edition), 9 October 1959.

[52] Saleh Salem in *Al Gumhuriya,* cited in *Bourse Egyptienne,* 10 Oct. 1959.

[53] *The Times,* 16 October 1959.

[54] See the report in *Mid East Mirror,* vol. 9 (1957), No. 22, pp. 13–14, 2 June 1957.

CHAPTER VIII

# Harnessing the Orange River

H. J. SIMONS

THE Prime Minister of South Africa, Mr. John Balthazar Vorster, pressed a button on November 18, 1966, at Oranjekrag on the Free State side of the river and set off 1,500 lbs. of explosives to mark the official start of work on the Hendrik Verwoerd dam. He was well qualified, he quipped, to trigger off explosions, having had five years' experience of those caused by his government's internal enemies. The dam was being built as an act of faith in the future of South Africa and a sign of determination to defend its internal borders against attacks from outside. In addition to the strategic and political aims, he referred to the project's social and economic benefits. It would add greatly to the country's food supplies and attract settlers to a vast and sparsely populated region.

The river rises at a height of 10,000 feet above sea level in the Drakensberg mountains of Lesotho, flows westward along a course of 1,300 miles through deep channels, mountainous tracts and arid country, traverses three-quarters of South Africa from east to west, and discharges in the Atlantic above Alexander bay. This great artery is the country's largest and most important river. It runs between highly developed industrial and agricultural complexes to the north and south; and forms, for several hundreds of miles, the boundary with South West Africa. Political, strategic and socio-economic considerations enter into the ambitious, long-range proposal to harness the river's waters and make them available for farming and industrial purposes.

### WASTED WATER

The Orange receives the waters of two other major river systems, the Caledon and the Vaal; drains two-thirds of South Africa's area, and almost the whole of its interior plateau; yet irrigates a very small area. Up to 10 million acre-feet of water flows annually down the Orange into the Atlantic, and carries with it many million tons of irreplaceable topsoil. It has long been realised that this drain on the poor natural resources of soil and water will if unchecked severely limit the possibilities of large, sustained economic growth. Engineers and geologists have discussed since the begining of the century ways and means of developing the Orange, or of diverting its waters to other basins. The present comprehensive project is the outcome of these investigations.

Nearly four million acre-feet, or two-fifths of the annual run-off, are derived from the Vaal. It is the only river system in South Africa that has been developed on a big scale. Its catchment serves the great industrial complex between the Orange Free State goldfields and Pretoria, and also the country's largest irrigation scheme in the Vaal-Harts valley. All the water from the Vaal will be required in its own catchment area and, consequently, should be disregarded in considering the prospects for developing the Orange. Indeed, it is expected that the project will eventually relieve the Vaal dam of the demands now made on it to provide water for the irrigation works along the middle course of the Orange.

Another million acre-feet, in round numbers, will be reserved for consumption in the Orange Free State and in the upper reaches of the catchment. This quantity must also be deducted from the annual run-off. A further deduction, amounting to less than a quarter million acre-feet, must be made in respect of the run-off derived from the huge expanse of country below the Vaal-Orange confluence, for this supply is too irregular to be used effectively.

The balance of about 4·5 million acre-feet, most of which comes from Lesotho, is available for the project. An allowance should be made, however, for the big losses suffered through evaporation and wastage during high floods over the spillways of the shallow, large-surfaced storage dams. When the losses have been taken into account, an estimated run-off of 3·4 million acre-feet will be available annually for effective use at a constant rate of 4,700 cubic feet a second.

## IRRIGATION POSSIBILITIES

It may seem, at first sight, that preference should be given to schemes for utlising this mass of water within the river's own basin. Indeed, some of the earliest attempts at irrigation in the Cape were made along the Orange below Upington, where an accumulation of silt has thrown up numerous islands. The Dutch

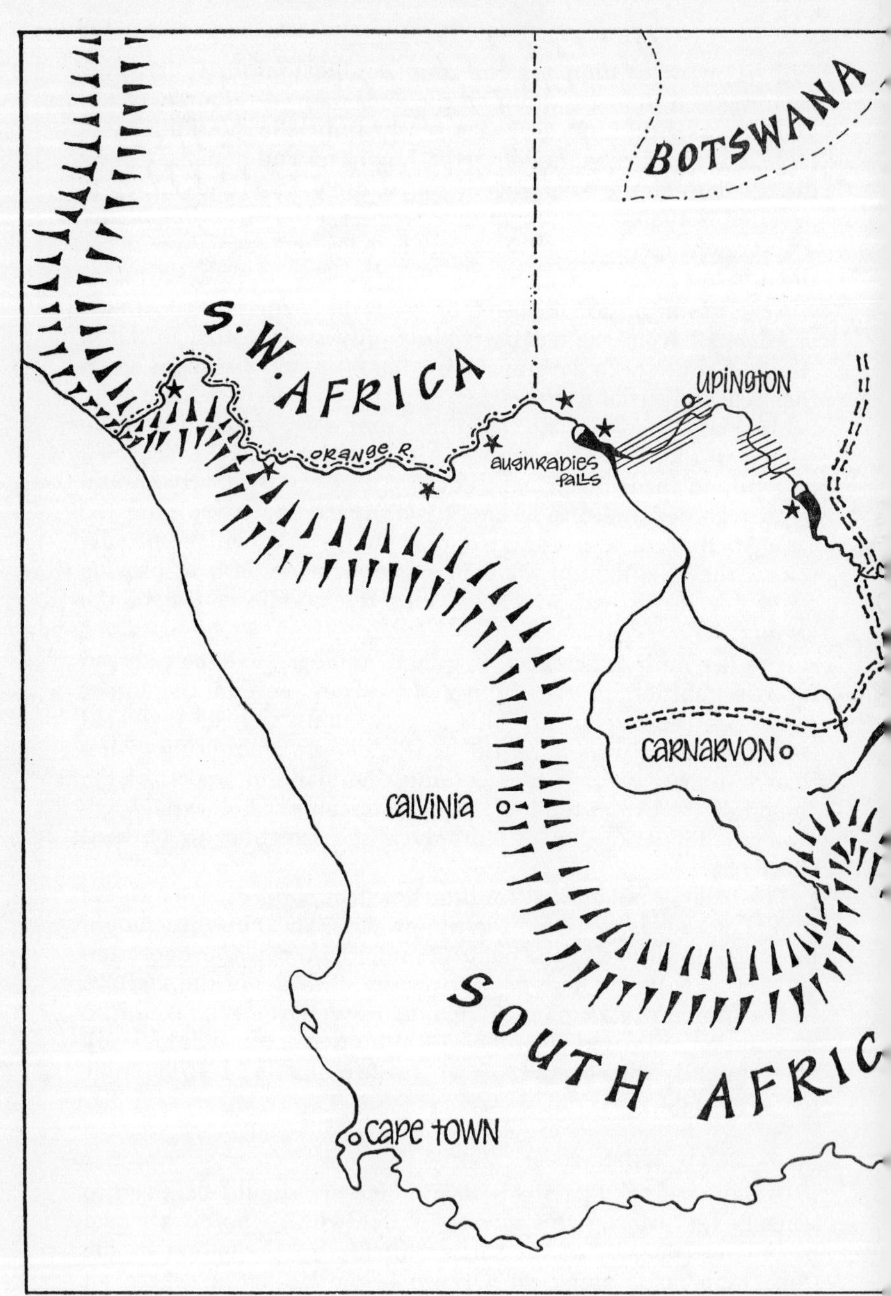


4. The Orange F

opment Project

Reformed Church founded a colony in 1898 at Kakamas for destitute farmers. Other flood irrigation settlements followed in this region. A storage reservoir, completed at Boegoeberg in 1934, supplies water, in years of low flow, for the 53,000 acres of land irrigated between the dam and the Aughrabies falls. Another 4,000 acres have been irrigated in the basin between the Falls and Alexander bay. But conditions have discouraged an extension of the irrigated land in the river's valley.

One handicap is the erratic nature of the rainfall, which occurs throughout the catchment in the summer months September to April. The river is flooded in years of heavy rains and ceases to flow in the lower valley after long dry periods. Then too, the Orange is deeply entrenched below the surface for most of its course, and runs through deep gorges from which the water would have to be pumped when the river is not in flood. There are few places on the upper reaches where a combination of good level ground and suitable dam sites would make irrigation feasible. River terraces with stretches of alluvial ground occur more frequently below the Vaal-Orange confluence, but the strips of alluvium are generally narrow and liable to flooding. For such reasons, it has been generally assumed that barely more than one-tenth of the river's water resources can be used profitably in its own valley.

If the water cannot be used in the basin, can it be diverted into other valleys with a greater irrigation potential? A. D. Lewis, the government director of water affairs, suggested in 1928 that part of the stream might be diverted through tunnels into the Great Fish river valley. T. B. Bowker, the member for Albany, took up the proposal and urged it on the Nationalist party government which took office in 1948. He received little encouragement. Successive ministers doubted whether it was practicable or necessary to bring water to the Fish and Sundays valleys. They already had three dams which would ensure water for the 36,000 morgen under irrigation for years to come. If the country went in for big schemes like the Orange-Fish project it might be faced with overproduction of foodstuffs. Tried and tested schemes, like that of Pongolapoort, should have preference. Bowker and the United Party opposition were accused of wanting to settle Africans on land irrigated with water from the Orange-Fish-Sundays project. The Minister of Water Affairs told farmers of the Eastern Cape in June 1959 to forget the tunnel scheme, which would cost £70 million, or nearly twice the amount estimated in 1951.

The government's policy changed dramatically after the Sharpeville massacre of 1960 and during the state of national emergency that followed. The minister announced in May that his government, under Dr. Verwoerd's leadership, would make a start with the Orange River Project before the next general election in 1963. Backbenchers were enthusiastic. The Cape, they said, was traditionally the white man's country, and the north-west Cape his hinterland. Inhabited by hardened, proud, freedom-loving farmers, and comparatively free of Africans, it held great possibilities of development and could not afford to lose its white population. Dr. Verwoerd declared in 1963 that the Orange River Project would recreate much of the southern portion of the Republic, put it on the map in many ways, and lead to the increase of the white population.

It was a case of defence before opulence. Ideological motives provide a major, if not the dominant, incentive. The scheme must be evaluated on strategic and political grounds, and not only in terms of economic advantages. But the initial impulse came from the settlers of the south-eastern Cape. To estimate the strength of their claims for relief we must look briefly at its river system.

It is dominated by two major rivers, the Great Fish and the Sundays. They rise along the great escarpment which constitutes the extremity of the interior plateau, cut through mountain ranges and traverse the dry Karroo, before reaching the coastal plains and discharging into the Indian ocean between Port Elizabeth and East London. Large stretches of rich alluvial soil fringe both rivers between the relief barriers and on the plain. Dams were built between 1918 and 1925 across the Sundays river at Lake Mentz and Van Ryneveld's Pass, and on the Brak and Tarka tributaries of the Great Fish at Lake Arthur and Grassridge, to irrigate land on which impoverished ostrich farmers and others could grow fruit and tobacco.

About £100 million have been spent on developing the valleys of the two rivers. £25 million have been invested in citrus plantations alone in the Sundays river valley, now a major fruit growing and exporting area. But the irrigation schemes along the rivers have not come up to expectations. Some authorities go so far as to say that they are in jeopardy because of erratic water supplies and a high rate of siltation. Much of the money spent on their development will be lost unless action is taken to ensure an abundant and regular flow of water. This is what the Orange River Project is intended to provide.

THE PROJECT

The central feature is the Hendrik Verwoerd dam, now under construction at a place three miles from Norvalspont below the inflow of the Kraai and Caledon. The dam is sited at a point from which it can be used to control the flow of the Orange and also serve areas both outside and within the watershed. Most of the run-off from Lesotho will enter the reservoir. It will be able to store a maximum of 5 million acre-feet of water behind an arched wall 280 feet high and 2,970 feet long. The length along the crest of the subsidiary earth embankment will be 5,000 feet. Water will be supplied under pressure to a hydro-electric power station erected at the dam. The release of water will be determined by daily and seasonal requirements and, it is estimated, can be so controlled as to reduce by half the occurrence of floods likely to damage cultivated land on the banks of the lower Orange.

A circular, concrete-lined tunnel, 16¾ feet in diameter and 51½ miles long, will direct water at the rate of 1,200 cusecs from the reservoir, at a point 26 miles above the wall, to the Thebus river near Steynsburg. Water will flow from there into Grassridge dam. This will be used to regulate, by means of canals, the water supply for areas along the Great Fish and Sundays rivers. It is estimated that the tunnel will supply supplementary irrigation water to 46,000 acres of existing development in the Fish and 23,000 acres in the lower Sundays river valleys; and that the combined supplements will have the effect of bringing 34,000 acres under new irrigation in the two valleys.

This development is envisaged for the first phase. The construction of more tunnels and canals in subsequent phases will provide water for the irrigation of an additional 137,000 acres in the Fish river valley and 67,000 acres in the Sundays river valley. Water will also be made available to Port Elizabeth and towns in the two valleys.

Turning now to the western flank of the main waterworks, we may trace the flow from the Hendrik Verwoerd dam through the turbines of the power station and down the Orange for 65 miles, to where it enters another dam, known as the Vanderkloof. This will be a diversion weir, 330 feet high, 2,600 feet long, and with a gross storage capacity of 2·5 million acre-feet. Its main purpose will be to feed water, by gravity or under pressure, to irrigable land and towns on both sides of the Orange.

A canal system, leading off from the left bank, and extending initially for 50 miles, will irrigate 32,000 acres on the left bank between Vanderkloof and Hopetown. A pump station on this bank will deliver a million gallons of water per day to De Aar, a centre of the Karoo's trade in sheep and cattle. It is contemplated that the canal may be extended, in later phases of the scheme, for a total length of more than 500 miles past Britstown and into the valley of the Sak river. The extension will enable another 170,000 acres to be brought under irrigation.

A second canal, taken out on the right bank at Vanderkloof, will serve 30,000 acres of irrigable riparian land in the Orange Free State, along the river above Hopetown, and in the Riet river valley, within the basin of the Vaal river. Power generated at the dam will be used to pump the water for the Riet river irrigation area. Eventually, according to the plan, additional canals and pumping stations will supply water for a further 78,000 acres of irrigated land in this region, and 20 million gallons a day to Kimberley and its environs.

The Orange Free State will benefit directly in a second way, from water pumped at the Hendrik Verwoerd dam, through 120 miles of pipeline, to irrigate 2,000 acres in the Kaffir river settlement, and, further north, to provide Bloemfontain with 15 million gallons a day. The piped water will be available also for towns and villages along the route.

A third high dam, projected for the upper Orange above its confluence with the Vaal, will be constructed in the second phase at Torquay with a capacity of more than 300,000 acre-feet. A gravity canal system leading from the dam will serve 57,000 acres of irrigable land near the confluence and further down the valley towards Prieska.

### THE IRRIGATED AREAS

The water flowing down the river below Prieska will be used, firstly, to free another 57,000 acres of existing irrigated land of dependence on supplies from the Vaal river. Secondly, 21,000 acres of new land will be brought under irrigation along the middle and lower Orange: at the Kakamas-Warmsand settlement (6,300 acres), the Marten and Krapohl islands (4,200 acres) and the Richtersveld-Alexander bay area (10,500 acres).

In summary, it is envisaged that the project, in the first phase, will provide water for 100,000 acres of existing and 75,000 acres of new irrigated land. The 175,000 acres will be distributed as follows: 15,000 along the lower Orange, between Aughrabies falls and the west coast; 63,000 on the middle Orange, between Boegoeberg and the falls; 32,000 in the Orange Free State, on the right bank of the river and in the Vaal basin; 32,000 on the left bank of the upper Orange; and 33,000 in the south-eastern Cape, in the valleys of the Fish and Sundays rivers.

The estimated capital cost of the water control works constructed in the first phase of the scheme is £65 million, exclusive of the cost of hydro-electric power installations. Cost estimates of the whole project, which is to be completed in six successive stages over a period of twenty or thirty years, can hardly be accurate. The tentative figure, as given, is £225 million, of which £150 million will be spent on water works, £45 million on hydro-electric power stations, and £30 million on buying and developing new land for settlement.

The completed project is expected to add an equivalent of 760,000 irrigated acres, which is an increase of 40 per cent in South Africa's total area of 1·9 million acres now under irrigation. It may not seem much, when set against the estimated capital cost of £240 an acre, or the annual gross value of £80 per acre, this being the anticipated yield of agricultural produce from the area to be irrigated. Experience has shown that water engineers, like other planners, tend to be over-optimistic when drawing up prospectuses to solicit support for their schemes. It is also notorious that the success of irrigation settlements in South Africa rarely comes up to expectations. Costs are likely to be greater, and yields smaller, than the estimates provide.

Important ancillary benefits are claimed for the scheme. It will diminish the incidence of damaging floods in the middle and lower Orange. Irrigation settlements now in jeopardy, as along the Sundays river, will be assured of a constant and abundant supply. Pressure will be taken off the Vaal catchment area, now strained to the limit of its capacity. A number of towns, potentially important centres of commerce and industry, will draw water from the Orange. The establishment of new settlements will help to reverse the flow of white persons to the towns, and attract them to rural areas that have been steadily losing their white residents for many years.

The taxpayer, who has to find the money for this expensive project, may reasonably ask whether a bigger return, in social and economic gain, could be expected if his money were invested in other undertakings. To make this kind of evaluation, it will be necessary to examine different aspects of the project in relation to the national economy and social policies.

POLICIES AND PROSPECTS

The generation of hydro-electric power is one of the great advantages claimed for the project. To meet the demand for power, especially in the north-western Cape, the government decided in 1964 to proceed immediately in the first phase to the construction of the Hendrik Verwoerd and Vanderkloof dams to their maximum planned height. This would allow for a fourfold increase in the power generating potential at the two dams during the project's first three phases. The Electricity Supply Commission considered that it would have to invest as much or even more capital in thermal power stations as the additional capital expenditure required to raise the dams to an elevation of 60 and 40 feet respectively above the heights initially proposed.

Power will be generated in the first two phases by using the heads created at the Hendrik Verwoerd, Vanderkloof and Torquay dams and the rush of water down the Aughrabies falls. More power will be generated in subsequent phases by installing turbines at the outlet of the Orange-Fish tunnel, at Boegoeberg dam, and at other points along the lower reaches of the Orange. An estimated potential output of 145 megawatts, rising to 197 megawatts in the final phase, will be obtained annually from the projected hydro-electric works. The generating stations, plant and transmission lines will cost an estimated £25 million. A like amount is to be spent on the special dams, canals and balancing reservoirs needed to develop the power sources.

It has not yet been shown that the power to be generated will be fully utilised or that other sources could not be developed more economically. South Africa produces large amounts of coal and uranium at relatively low cost. The capital cost of conventional coal-fired or nuclear power stations may be much less than the cost of the hydro-electric power stations. Nuclear and to a lesser extent coal-fired stations have the additional advantage that they can be sited close to the main consuming centres. Hydro-electric power may be a valuable by-product but its provision can hardly be said to constitute a strong argument for the Orange River Project.

A major consideration in its favour is the urgency of the need to supplement the Vaal river's resources. They will soon be inadequate for the needs of mining, industrial, farming and domestic consumers in the catchment area. The recent prolonged drought has driven the point home. The water level in the Vaal dam was still falling at the end of 1965, after a third of the highveld rainy season had passed. Some authorities doubt whether the project can be developed in time to save the Witwatersrand complex from a water famine. They claim that the additional supplies required can be obtained sooner and at lower cost by diverting Lesotho streams into tributaries of the Vaal.

As has been pointed out, the Orange river's main headwaters rise near the edge of the Drakensberg escarpment in Lesotho. The mean daily supply of water is said to range from 130 million gallons at an elevation of 8,000 feet to 800 millions at 6,000 feet. Silting and evaporation would be relatively small in the upper catchment areas. Water could be gravitated over a distance of 200 miles through the Free State and as far as Kimberley and the Vaal dam. One possibility is to dam the northernmost headstream, the Madimamatso, in the Maluti mountains at Oxbow, or lower down at Kau. It is estimated that if the Oxbow scheme were adopted some 20 million gallons could be delivered daily to the Caledon river at small cost. The Rand Water Board could augment its supplies to the extent of 30 per cent by drawing on the Oxbow catchment area, and 75 per cent by gaining access to the Kau catchment.

South Africans have debated these possibilities for many years with inconclusive results. The Sotho want to sell their water. It is potentially their most valuable export apart from the manpower they supply to labour centres in South Africa. South Africa's main objection seems to be political and strategic. Lesotho, it has been said, could hold South Africa to ransom by threatening to blow up dams, poison the streams, or sabotage the supplies in other ways. For these reasons, or because of reluctance to depend on what is constitutionally a foreign state, the government has preferred to tap Lesotho's streams after they have entered the Republic.

### AGRICULTURAL GAINS

The Orange River Project, as conceived, stands or falls by the value of its contribution to land utilisation and agricultural output. To estimate these, we need to look at the present state of farming in South Africa. White farmers own 226 million acres, or about 75 per cent of the surface area. 25 million acres are cultivated, as compared with 18 million in 1950. An estimated 19 million acres are under the most important cereals, oilseeds and sugar cane.

An additional 760,000 acres of irrigated land, amounting to 3 per cent of the existing cultivated area, can hardly make a significant difference to the country's food supply in normal circumstances. The recent severe drought, which afflicted the maize belt for several years, has shown that South Africa is not insulated against extraordinary mishaps. But the price of achieving absolute security may be excessively high. Moreover, it has not been shown that expensive irrigation works are necessary to bring about the desired increase in output.

Dryland farming is cheaper than producing the same crop under irrigation. It is usually uneconomic to grow low-priced crops with a low margin of profit, like cotton, groundnuts, maize and lucerne on irrigated land. Peasant cultivators, accustomed to a low standard of living, may find this culture satisfactory. White South Africans, under similar conditions, are usually subsidised by the state to an extent of 80 per cent on government schemes.

Farmers have steadily raised the productivity of dry lands by means of mechanization, pest control, and expanded use of fertilisers and improved seed strains. The volume of agricultural production increased by an average of 3·7 per cent a year from 1947/8 to 1962/3, while the total food consumption increased by only 2·3 per cent. Overproduction, rather than scarcity, has been the major agricultural problem for several years. Past performances suggest that dryland farming, with improved techniques, could ensure an adequate supply of foodstuffs for the forseeable future.

Maize is a case in point. About one-third of all white farmers grow maize for the market. They cultivate 10 million acres, which is 40 per cent of the entire cultivated area, and average 6 bags per acre as compared with 3 bags before the second world war. The average yield could be raised to 10 bags under dryland conditions, resulting in an output of 100 million bags. The present output is about half this amount, and exceeds local annual consumption by an average of 26 million bags.

The 760,000 acres of newly irrigated land are expected to yield products of a gross value of £57 million a year, representing 13 per cent of the total value of agricultural output, and increasing the net national income by 1·6 per cent. Such long-range calculations, projected over a quarter of a century, cannot but be highly speculative. They are ideological targets, and are intended to justify the expenditure of large sums of public money on projects that serve rather different purposes than those indicated by the estimates. It is significant that they are not presented in the form of a comparison with possible alternatives, based on existing trends, such as probable yields from comparable investments in dryland farming or other sectors of the economy.

Given an adequate supply of basic foodstuffs and raw materials, South Africans may do better for themselves by investing in secondary industry, or in providing the infra-structure required for industrial growth. The country is richly endowed in mineral resources, but lacks water and power in a number of centres capable of being developed industrially. Of the 3·4 million acre-feet of water to be derived annually from the project, only 190,000 will be made available to industrial and urban consumers. This imbalance reflects a scale of preferences in which priority is given to specific social factors at the expense of economic growth.

THE RURAL EXODUS

One of the factors concerns an alleged 'depopulation' of the rural areas. In 1921 half the white population lived on the platteland, a term describing farms and small country towns. The proportion is now hardly more than one-quarter. Only 18 per cent of the whites reside in the designated rural areas, many of which are virtually extensions of an urban complex, such as the Western Cape, the Orange Free State goldfields, and the Witwatersrand. Some 15,000 farms were found in 1954 to be occupied exclusively by African, Coloured or Indian employees. They worked the farms on behalf of absentee owners, or under the supervision of a white person who lived elsewhere. Most of these farms were situated in Natal and the Orange Free State.

The townward 'drift' is the result of a rapid industrialisation. It has been mainly instrumental for the substantial improvement in the living standards of South Africans. The strong rural exodus that set in after 1910 provided the solution of the 'poor white' problem, once a running sore in the body politic. More than half the number of gainfully occupied whites are employed in the tertiary sector. This indicates a high level of economic maturity. It could not have been attained without a shift of population to the urban centres.

While recognising the benefits of industrialisation, a dominant section of the ruling élite disapproves strongly of the persistent movement to the towns. The influential Carnegie Commission on the poor white problem urged, in the early 1930s, that:

> 'It is a question of national importance whether the rural population, with certain valuable qualities, is not undergoing a too great decrease; particularly since they are being replaced in many districts by natives. And in any case the rural exodus is a serious indication of social and economic malady in South African farming and farm life'.

The commission of 1956–9 on the rural white population expressed similar views in stronger terms. The 'White platteland', it argued, 'is largely the pivot of Western civilisation'; a strong farming community 'is a prerequisite to the continued existence of Christian civilisation in our country'. White civilisation in the cities would not be able to hold its own if the pivot were to collapse. A great danger arises from the 'enormous increase in non-whites on the farms'. If present trends continue, 'the country's basic material need—its food production—would ultimately be in the hands of non-whites'. If this were to happen, 'civilisation would automatically have to capitulate' to those who produce the food of the nation.

White supremacy is here equated with Western civilisation. The adoption of measures to reverse the movement of whites, and to make drastic reductions in the number of Africans on white-owned farms, is presented as a major strategic aim in the effort to maintain white supremacy. It is known that the Commission's findings made a deep impression on government leaders, and appreciably strengthened the case of those who argued in favour of the comprehensive Orange River Project.

Yet the project cannot be expected to have a significant effect on the distribution of population. Not more than 9,500 white farming families, a total of 45,000 persons, could be settled on the newly irrigated land if it were divided into farm units averaging 80 acres each. This process, it should be remembered, is to spread over thirty years. But whites have been abandoning the rural areas at a probable rate of 8,000 a year since 1931. Far from stemming the flow, the project may well accelerate it by increasing the water supplies of metropolitan regions.

### BLACK AND BROWN SOUTH AFRICANS

The 'rural exodus' is deplored because it adds to the relative strength of the African and Coloured farm population. It is not unreasonable to say that the project originates largely in policies concerned with race relations, and the role of the Afro-Coloured group in the social structure. Yet they receive only a passing reference, if any, in the discussions. The whole project falls within the 'white group' area, except for a stretch of the lower Orange in the Kalahari; and the function of Africans and Coloured persons in the 'white man's country' is to provide manual labour for a white employer. Their benefits from the scheme are therefore assessed mainly in terms of employment opportunities created on construction works and irrigated lands.

The Afro-Coloured group will be affected, however, in a much more profound way. For one thing, the investment of large sums of public money in the Orange River basin must divert resources away from the development of the 'border' regions of the African reserves, called 'homelands' in current official phraseology. Secondly, the project is bound to retard the development of the homelands themselves. £57 million were to be spent on them in the five years 1962–6. Sixty seven per cent of this amount would go into the building of houses. The balance was to be spent on roads, irrigation, water supplies, afforestation and public buildings. These projects may improve agricultural productivity. More rational farming methods will, however, increase, not reduce, the surplus population. It can be absorbed in the reserves only if they are industrially developed. This is hardly possible while the men and material required for the purpose are employed on the Orange River Project.

It will attract peasants from the reserves in two ways. Some will be employed on the construction of dams, tunnels, canals and subsidiary works. Secondly, it will stimulate development in the existing industrial areas, and so increase the demand for African labourers. For these reasons, it would seem that the project will intensify the drain on the manpower of the reserves and cause them to lag still further behind the rate of economic growth in more developed regions.

By concentrating its resources on the white sector, the government has found a ready reply to critics who accuse it of favouring the African at the expense of the taxpayer and the farmer. This may be a gain in terms of party politics. But the diversion of resources will strengthen the suspicion that 'separate development' is a euphemism for 'white domination'.

The prospects of the Coloured inhabitants of the lower Orange area and the south western Orange Free State may be somewhat brighter, where they are in a position to benefit directly from the irrigable lands. The most important of these rural communities are in Namaqualand and the north western Cape. Some 18,000 persons live here in vast reserves whose combined area is nearly 4 million acres. Most of the region receives between 4 and 10 inches of rain a year. The vegetation consists of widely spaced Karoo bushes and fine tufted grass. Areas of bare shale and sandstone are common. Trees are rare, except along the river beds. The inhabitants graze sheep, but depend mainly on wages earned on farms and in copper mines and fish factories in the white sector. This arid region, where farmers 'farm stones', according to the local expression, has been declared the 'homeland' of the 1·5 million Coloured people.

Some of the Coloured reserves fringe on the Orange. Irrigable riparian land in this cluster will receive a constant supply of water and protection against flood when the project has been completed. Only about 10,000 acres, situated in the Richtersveld will however be irrigated specifically for the Coloured, some 2,500 of whom are to be settled on 20 acre plots. White settlers, in contrast, will have access to 65,000 acres of existing and irrigated land along the middle and lower Orange. The discrimination corresponds to the traditional racial pattern. But its perpetuation means that the administration is once again foregoing an opportunity to inject reality into the concept of separate racial development.

Namaqualand is the ancient home of the Nama, one of the main ancestral groups from whom the Coloured are descended. They can claim rights of first occupation. But this priority has not protected them against white encroachment in the past, nor will it necessarily protect them in days to come. A government committee was appointed in 1951 to consider whether Coloured should be replaced by whites at Eksteenkuil and Witbank, two stretches of irrigated land in and along the Orange. The committee was also instructed to consider the introduction of racial group areas, and the possibility of removing the Coloured to other land along the Orange. The trend of opinion in ruling circles has been more favourable to them since Sharpville. Stress is being laid on the need to cultivate their goodwill and loyalty. But it is not suggested that they should be allowed to benefit equally with the whites from the Orange River Project.

Diamonds, tungsten, beryllium, quartzite and other minerals occur in the Nama reserves, which are adjacent to the diamond coast above Alexander Bay. Concessions for prospecting and developing have been granted to the Coloured Development Corporation, a statutory state organization, or to private white companies and individuals. No individual Coloured applicants have been successful, on the alleged ground that they lack the necessary capital and technical skills. The only exception is in the Leliefontein reserve, where the corporation has agreed to assist a company of Coloured diggers. It is doubtful whether the Coloured residents of the region will be able to hold their own on valuable land against the pressure of white entrepreneurs and settlers.

The Orange river forms, as has been pointed out, the boundary between South Africa and South West Africa below the Aughrabies falls. The status and future of the mandated territory are contentious and potentially explosive international issues. Some authorities assert that the projected development along the upper and middle reaches of the river will diminish the flow of water below the falls and adversely affect climatic conditions on both sides of the river. If this prediction proves to be correct, the national liberation movement in South West Africa may be expected to become hostile to the enterprise. In the event of a clash, the north-western Cape is likely to be an important strategic zone. Intensive white settlement along the river may then serve the same purpose as Sir George Grey's buffer settlement of legionaries between the Keiskamma and Kei rivers in 1857. This possibility was probably also taken into account when the policy makers decided to adopt the comprehensive Orange River Project.

# BIBLIOGRAPHY

R. E. Altona, 'Agriculture in the Republic of South Africa', *South African Journal of Science,* 59, 1963, p. 517.

T. B. Bowker, 'Claiming the Boundry of the Orange River', *Optima,* 13, 1963, pp. 27–9.

Carnegie Commission, '*The Poor White Problem in South Africa,* Vol. I, 1932.

Monica M. Cole, *South Africa,* 1961.

*Construction in Southern Africa,* Vol. 10, No. 3, 1965, pp. 55–63.

Department of Water Affairs, 'An Old, Old Dream becomes real', *Farming in South Africa,* 39, pp. 6–22.

T. J. D. Fair, 'Regional Aspects of the O.R.D.P.', *South African Journal of Science,* 59, 1963, pp. 455–460.

J. J. Heynike et al., 'Optimum utilization for Industrial Use under the O.R.D.P.', *South Africa Journal of Science,* 3, 1965.

*Orange River Development Project and Progress Report,* 1965.

J. Phillips, Aspects of the Ecological Approach etc.', *South Africa Journal of Science,* 61, 1965.

C. P. Robinson, 'How the O.R.P. will work', *Optima,* 13, pp. 30–9.

J. J. L. Stallebras, 'The O.R.D.P.', *South Africa Journal of Science,* 59, 1963, pp. 448–454.

T. Van Waasdijk, 'Some Secondary Economic Effects of the Execution of the O.R.P.', *South Africa Journal of Science,* 59, 1963, pp. 468–472.

South Africa, *House of Assembly Debates,* 1951–1965.

........................, *Coloured Affairs Department Annual Reports,* 1952–1964.

........................, *Report of the Commission of Inquiry into European Occupancy of Rural Areas,* 1960.

........................, *Secretary for Water Affairs, Report on the Proposed O.R.D.P.,* 1962.

........................, *Report of Secretary for Water Affairs for 1963–4,* R.P. 18/65.

........................, *Department of Agricultural Economics and Marketing. Annual Report for 1963–4,* R.P. 13/65.

CHAPTER IX

# International Legal Aspects of the Kariba Project

R. H. F. AUSTIN

ALTHOUGH originally conceived as a purely national project, and constructed in the heart of the Federation of Rhodesia and Nyasaland to provide a functional reinforcement for the somewhat unsteady political structure of that experiment, Kariba owed some part of its existence to international legal institutions in the form of a substantial loan from the International Bank for Reconstruction and Development (hereafter referred to as the World Bank)[1]. The loan itself was made to the Federal Power Board, and guaranteed by the United Kingdom and the Federation. Since 1956 however, the Kariba project has moved further and further into the international legal sphere, as first the Federation of Rhodesia and Nyasaland was dissolved[2] at the end of December 1963, and later, on the 24th October 1964 Northern Rhodesia became the independent state of Zambia. Thus, the highly ambitious, and enormously expensive project, born within the framework of the municipal legal system of a single dependent State, under the overall direction and control of the United Kingdom, must needs be operated within the much less integrated system of international law, and against a political background (at least since the Unilateral Declaration of Independence by the former government of the colony of Southern Rhodesia), which makes functional cooperation between the two entities most directly concerned extremely difficult[3]. Before examining some of the problems raised by Kariba and the machinery created to deal with them, it is necessary to set out briefly what the project involves.

## THE KARIBA SCHEME

Like so many ventures concerned with the development and exploitation of water resources, Kariba is more than a dam holding back the waters of a particular river: it is an organic scheme, intended to grow and to play an active part in the development of the two territories to its North and South. The loan made by the World Bank in 1956 was in respect of 'the programme for the production of power by the installation of a Hydro-electric plant, having a maximum capacity of 1,200,000 kw, in the Kariba Gorge of the Zambesi River, and the installation of related facilities for the supply of such power to major consuming centres in the Federation through an interconnected system of transmission'[4]. It is immediately apparent from this provision that the consequences of the scheme were intended to extend deep into the territories of both Rhodesias as they then were, the significance of which increased with their constitutional movement toward sovereign independent status. The first stage of the scheme involved the building of the dam wall and the installation of a power plant generating approximately 600,000 kilowatts on the South bank of the river, together with the erection of power lines stretching from Bulawayo in the South to Kitwe in the North. The scheme was vitally important to the development of the Federal economy, particularly to the expanding copper mining industry in Northern Rhodesia. In 1960 the Monckton Commission suggested that 'Without the advantages of central planning and execution, based upon a market capable of absorbing more and more power, its seems unlikely that such an ambitious project could have been financed and undertaken', and the question since the dissolution of the Federation has been whether this 'most spectacular achievement'[5] could continue to exist and achieve its full potential. The economic integration achieved during the Federation could not be unscrambled in a day, particularly since the two most vital utilities, Transport and Power were involved.

The forms of international functional cooperation most frequently discussed are, quite understandably, the positive efforts being made throughout the world, but especially in Europe, to achieve economic integration and cooperation followed in some cases by political developments along the same lines. In the context of Central Africa, and in other areas emerging from dependence[6] a similar process is going on, albeit of a somewhat negative variety. The atmosphere of recently acquired sovereignty is apparently not conducive to the growth of institutions calling for the limitation of sovereignty for the achievement of effective means of dealing with the tremendous tasks of economic development and reconstruction facing former colonial areas. Yet Kariba has continued to grow as planned, despite the sharp political differences between the governments of Zambia and Southern Rhodesia, a clear demonstration of the strength of such functional ties. At the break-up of the Federation in 1963, the two territories expressed the 'desire that the integrated system for the control of the generation of electric power and its transmission in the territories . . . should continue to be operated and fully developed as a single system under the joint ownership or control of the . . . Governments'[7]. There followed the enactment of legislation by all three governments concerned[8], and fresh agreements with the World Bank[9]. In terms of this legislation the Central African Power Corporation was established for the two territories to continue the work of the former Federal Power Board, under the overall control of the Higher Authority for Power on which the two territorial governments enjoy equal representation. Within a year of the dissolution of the Federation, the arrangement proved its feasibility as the second stage of the scheme proceeded, with the assistance of another loan, of $7·7 million, made by the World Bank to the new jointly owned Power Corporation[10]. Despite the political uncertainties surrounding Kariba at the time, it is interesting to note that ten private institutional investors, mostly United States bodies, agreed to participate in the loan, without the World Bank's guarantee, to the extent of $4,540,000[11]. The estimated cost of the second stage work being undertaken is $15·4 million, half of which will be met by the Corporation out of its own resources, without any call upon the taxpayers of either Southern Rhodesia or Zambia[12]. Further, it should be noted that the work being undertaken involves the construction of an additional power line to the Zambian Copperbelt all of it being within the boundaries of Zambia. The next part of the second stage is the installation of a 900MW power station on the North bank of the Zambesi which will end the present situation whereby all the actual generation of power takes place within the territory of Southern Rhodesia, giving the government of the latter an extremely powerful weapon to use against Zambia, if it were inclined to breach its obligations. Before examining the legal framework within which Kariba now operates, it is necessary to consider briefly the more general problem of joint international exploitation of water resources under international law.

### INTERNATIONAL LAW AND 'INTERNATIONAL RIVERS'

The sources of rules of international law in this sphere, as in all other spheres of international law, are exclusively those set out in Article 38 of the Statute of the International Court of Justice, namely: international customary law, international conventions and general principles of law recognised by civilised nations[13]. The governing principle most relevant to the problems involved in the exploitation of waters common to two or more States, is Sovereignty, against the background of which States may agree to limitations of their exclusive control, in favour of co-operation. Over the past century, States, especially those within the confines of Europe, have come to realise that the multitude of uses to which common water resources might be put, has created the need for some form of international regulation of such use[14]. Yet, on the level of international customary law, the principle of sovereignty remains dominant in the sense that joint efforts to develop 'international rivers' must be based upon the consent of the States within whose territorial sovereignty part of the resources falls[15]. As recently as 1952 a comprehensive study of the problems arising in connection with hydro-electric development had to conclude that 'while it cannot be said categorically that riparian States have clearly defined legal rights with regard to the hydro-electric development of rivers which are only partially under their sovereignty, all the riparian States firmly believe that such development work can be carried out only on the basis of their prior agreement'[16]. The learned rapporteur suggested that there was some evidence to suggest that Europe was emerging from an era dominated by the idea of rivalry between nations to a new era dominated by the idea of conscious solidarity, and propounded the idea of an 'international public utility' growing within an increasingly integrated Europe. But even in Europe with well over 150 years of developing public law concerning the use of waters common to more than one state[17], it is clear that consent remains the basis of international regulation or exploitation, a situation emphasised by the years of non-cooperation on the Danube between the European Blocs. Unless consent is forthcoming the State wishing to develop common resources may therefore face 'the tragedy that what is technically possible cannot be brought to function on the legal plane because inhibitions of a traditional nature may prevent the carrying out of what is both possible and reasonable'[18].

International ventures of the nature of Kariba must therefore find a basis in international conventional law, and it is here that the legal framework of Kariba is most interesting. The existence of Kariba as a going concern, it is clear, owes almost everything to the temporary political and economic integration which was the

Federation. Given their present political position it is unlikely that Zambia and Southern Rhodesia could have agreed to the creation of the project, no matter how inter-dependent their two economies might have been. Part of the justification for the scheme was the fact that it would decrease the Copperbelt's dependence upon Rhodesian coal, which it did as regards the thermal production of electricity for the copper mines; though strategically, the erection of the power plant on the Southern bank of the Zambesi has (as the events since November 1965 have proved) put the whole of Zambian economy at the mercy of the men who physically control that area. The legal structure set up to replace the Federal Government is clearly incapable of dealing with the political problems which threaten the utility of this vast scheme, and may reduce it to the 'Welensky's Folly' status predicted by some of its critics. Unless the current crisis involving Southern Rhodesia and Zambia (in which the United Kingdom and the United Nations are inextricably concerned) is resolved in such a way as to set these two communities on something approaching a parallel political course, the prospects of this African venture in functional cooperation would seem dim. Having made that observation on the all important political background of the experiment, it is necessary to examine the legal framework within which Kariba exists at present.

### THE FEDERAL DISSOLUTION AGREEMENT AND THE CENTRAL AFRICAN POWER CORPORATION

It has already been stated how, with the dissolution of the Rhodesian Federation by the United Kingdom Parliament, a separate Corporation was created to take over the functions of the former Federal Power Board[19]. A the same time the two territories stated their desire to continue the operation as a single system, and established machinery to achieve this end[20]. At that stage only the United Kingdom enjoyed full international personality. The agreement of the 25th November 1963 between the two Rhodesias, as they then were, was however sanctioned by the United Kingdom by provision in the dissolution Order[21]. At the same time the other entity concerned with the project, the World Bank, also brought its legal relationship into line with the changed constitutional situation[22]. In a sense Kariba was not at that stage an 'international' project, as both territories were United Kingdom dependencies. Yet there is no doubt that they enjoyed some degree of international personality, which in the case of Southern Rhodesia in particular, approached very close to independence.

When, within one year of the making of the agreement, Zambia achieved her independence, no fresh agreement was made, and both parties and the United Kingdom continued to act upon the assumption that Zambia's right to receive power produced on the South bank still bound the Southern Rhodesian government. The application of the *'inter se'* doctrine to such an agreement between States (dependent or independent) which are members of the Commonwealth may cause some doubt as to the exact legal nature of the agreement, and the law governing it. Yet, it is submitted that here is a clear case where the intention of the parties indicates an agreement governed by international law[23]. This view has been confirmed somewhat dramatically by the events following the Unilateral Declaration of Independence, the Prime Minister of the United Kingdom having referred to Kariba on a number of occasions as 'this international project',[24] and apparently recognising Britain's responsibility to ensure the flow of power from Kariba. In addition to this, the leader of the purportedly independent State of 'Rhodesia' has gone out of his way to make it clear to Zambia that his government 'has every intention of honouring its obligations especially in relation to the railways, the airways and the Central African Power Corporation which serve both our countries, in all circumstances'.

On the other hand it must be noted that the inter-governmental agreement has not been registered with the United Nations, which would be consistent with the non-sovereign status of the parties, and United Kingdom practice in such cases. There is another interesting feature of the legal arrangements relating to Kariba, in that the Guarantee Agreements made between the World Bank and the three governments involved, laid a special obligation upon those governments in their separate agreements with the Bank to 'perform, observe and be bound by all the provisions of the (November Agreement)'[25]. Thus the agreement is bolstered by the wider order provided by the World Bank's involvement in the project. It is submitted that in any dispute between the two governments which might arise for settlement by an Arbitrator appointed in terms of Article 20 of the November Agreement, the law applicable, unless otherwise agreed, would be international law.

The substance of the régime created by the two governments and the United Kingdom, relates primarily to the division of responsibilities already incurred, and the machinery to continue the scheme's operation. Most important of the latter is the Corporation. This is a creature of United Kingdom legislation, and is vested with powers and functions by both territorial governments to do everything which would otherwise be done by national producers. Consistent with other provisions in the régime, equality between the parties is maintained on the Corporation Board which is intended to be an expert body rather than a political one. Consultation between the parties is required to ensure that as far as possible this pattern is maintained. The Corporation may make majority decisions[26] in carrying out its functions, which are generally to supply electricity to undertakings in Zambia and Southern Rhodesia, and for that purpose to continue to operate the Kariba system; to establish additional undertakings for bulk supply as directed by the Higher Authority and to investigate additional facilities and advise the Higher Authority[27].

The Corporation then is intended to be the purely functional branch of the scheme, under the guidance of the executive and political organ which is the Higher Authority. Its practice indicates that it has been able to carry out these functions without much difficulty. Most impressive confirmation of this is the fact that the first part of the Second Stage, the construction of the new transmission line to Zambia, has been completed since its establishment, financed in part by a loan from the World Bank. The remaining cost of the project is met by the Corporation, which is bound under its agreement with the World Bank to follow sound business practices, and to be a 'financially self-sustaining public utility enterprise by earning revenue sufficient to cover all costs, to repay loans, and to pay for not less than half the development'[28]. To date the Corporation has succeeded in carrying out its obligations and better. It is a truly international public utility. The benefit of serving a consistently large market has actually led to a progressive reduction in cost of power. It is also of interest to note the Corporation's growing consciousness of its 'international responsibility'[29], which led it to take every precaution to avoid identification with one particular national group. This growth of 'internationalism' might be expected to develop even further if the institution continues to function as intended.

The November Agreement, while proving an adequate machine for the functional objectives of the parties, did not attempt to create an institution of a supra-national nature. Political control of the operation was maintained by the establishment of the Higher Authority, upon which both parties are equally represented, and whose directions must be unanimous to be effective[30].

The Corporation is bound to give effect to the decisions of the Higher Authority, which are to be 'of a general character' and made only after consultation with the Corporation[31]. In fact no 'directions' have yet been given by the Higher Authority. Apart from this overall control maintained by the parties, the Agreement contains specific undertakings by both, which place a considerable limitation upon their sovereignty. In the first place there is the general undertaking 'to assist the Corporation to carry out Stage II of the Kariba project progressively and in pace with the growth of the demand for power . . . and to give a proper place to this in the economic development programmes of their respective Territories'[32]. In addition attempts to frustrate the progressive development of the project are to be avoided by limitations placed upon the parties' freedom to develop purely national sources of power[33]. This limitation takes the form of compelling each party to consult fully with the Corporation before approving any national undertakings wishing to decrease or increase their production. Though the final decision is left in the hands of the particular Minister, the machinery ensures that the interests of the joint venture are fully considered.

Finally, the Agreement obliges the Higher Authority to look to the future development of the project, and the prospects of meeting future demands on the system, by requiring it to consider an annual Development plan prepared by the Corporation[34]. If the Corporation's plans should at any time fail to be accepted by the Higher Authority, and the demand should exceed the capacity of the generators, the Higher Authority is left to decide on the division of available power, paying special regard to the equal partnership of the two Territories. The principle of equality is in fact basic to the Agreement, certainly at the level where any conflict between functional objectives and political interests might occur. Yet the instruments upon which the joint venture is founded contain provisions on arbitration which imply a limitation of sovereignty beyond those already observed. This is found in both the Order-in-Council and the November Agreement[35], and is remarkable for the fact that it appears to be designed to prevent a deadlock in any situation where the Higher Authority is 'unable to render a decision or finding as to any question or matter necessary to the exercise of the powers or duties assigned to it', by *obliging* the Higher Authority to submit it to arbitration and appoint an arbitrator. The World Bank is given a specific role in this connection, being empowered to appoint an arbitrator should there be disagreement between the parties[36]. The award of the Arbitrator is binding upon the Higher Authority, and this particular provision of the Agreement applies *mutatis mutandis* to any disagreement arising between the two governments regarding the Agreement[37].

It is submitted that these provisions designed to ensure that the project will not fail through a need of executive guidance, arise from the fact that the future development of the project was envisaged within the somewhat integrated political structure of the Commonwealth, overlaid by the quasi-order[38] of the World Bank agreements. It remains to be seen whether this experiment in functionalism will survive the undermining of the former element, threatened by current events.

### THE LEGAL ORDER OF THE WORLD BANK AGREEMENTS

The various engagements entered into between the United Kingdom, the two Territories and the World Bank since 1956, provide an important set of regulations within which the common waters of the Zambesi River have been exploited. The examples provided by this structure and the case of the Indus Treaty, suggest that here, within the Bretton Woods framework, might lie a new and most useful instrument for the solution of legal difficulties facing would be developers of common water resources. Not only do the Loan and Guarantee agreements provide for the purely financial obligations of the parties concerned, but also for more positive activities. Thus in this case, the parties undertake to cooperate fully to ensure that the purposes of the loan will be accomplished, to consult with one another and with the World Bank, on matters relating thereto. The World Bank becomes, in fact, an independent overseer of the international project, the parties being bound to 'afford all reasonable opportunity for accredited representatives of the Bank to visit any part of the Territories . . . for purposes related to the loan'[39]. This power was given dramatic emphasis when the World Bank representative visited the Kariba Dam at a time when rumours that it had been mined by Southern Rhodesian rebels were rife. Indeed the World Bank is allowed a say in the overall economic planning of the borrower during the loan period, and is specifically to be consulted and must agree to any fresh undertaking which the borrower might contemplate prior to the completion of the project[40]. In addition the agreements provide in detail for any doubts concerning the overriding nature of the international obligations, as well as for arbitration in case of dispute[41].

What has the Kariba project to teach us? Within the order provided by these two agreements a *de facto* international public utility has been created and may continue to grow and serve. For the international lawyer seeking solutions to the problems of exploitation of common natural resources, and means of rescuing the economic benefits of central imperial control threatened by the potential 'Balkanisation' of de-colonisation, it highlights both the possibilities and difficulties. The machinery for international economic development conceived at Bretton Woods may be able to play a significant part in reducing the difficulties attendant upon the sudden move from dependence to independence, providing a legal framework less crude than that of international customary law. On the other hand little can be achieved without a minimum degree of socio-political parallelism between States, whether they be new or old. The quasi-order of the United Nations is perhaps intended to provide this, yet its history shows basic differences between its members are by no means uncommon, and the principle of good-neighbourliness[42] insufficiently developed to have any great significance[43]. But assuming that a minimum degree of political agreement is established in the area served by Kariba, or in other areas potentially capable of development along similar lines, the pattern evolved in the international agreements created to perpetuate the utility of this experiment may well serve future planners in this sphere.

[1] Loan Agreement between I.B.R.D. and Federal Power Board dated June 21, 1956. Loan Number 145 RN. Preamble. I.B.R.D. Press Release 448 of June 21, 1956. In terms of the agreement the World Bank provided $80 million for the first stage of the scheme, estimated to cost a total of $225 million (£80 million). At the time this was the largest loan made by the Bank for a single project.

[2] S.I. 2085 of 1963. The Federation of Rhodesia and Nyasaland (Dissolution) Order in Council. Part III ss.33 of 46.

[3] The disintegration of the centralised Air and Rail services, which do not enjoy the physical advantage making it virtually impossible to 'break-up' Kariba have already reached an advanced stage.

[4] I.R.B.D. Loan No. 145 Art. 1.02 and Schedule 2.

[5] 'Legacy of Progress'. Report to the Federal Assembly on 9th December 1963 by the Prime Minister.

[6] For a full examination of the problem in East Africa, with special reference to the fate of the Common Services, see T. M. Frank, *East African Unity through Law,* 1964.

[7] Agreement between the Government of Southern Rhodesia and the Government of Northern Rhodesia relating to the Central African Power Corporation, November 25, 1963. Northern Rhodesia Government Notice No. 2333 of 1963. Preamble.

[8] S.I. 2085/1963 (see note 2 above), Central African Power Act; No. 62 of 1963 (Southern Rhodesia), Central African Power Ordinance, No. 6 of 1964 (Northern Rhodesia).

[9] Loan Assumption Agreement (Kariba Project) between I.B.R.D. and Central African Power Corporation and United Kingdom, dated December 30, 1963. Guarantee Agreements between Northern Rhodesia, I.B.R.D. and United Kingdom, and between Southern Rhodesia, I.B.R.D. and United Kingdom, both dated December 30, 1963.

[10] Loan Agreement (Second Kariba Project) between I.R.B.D. and Central African Power Corporation, dated October 2, 1964. Loan No. 392 RNS. Also the Guarantee Agreements of the same date between the World Bank and each of the interested governments—United Kingdom, Southern Rhodesia and Northern Rhodesia.

[11] I.R.B.D. Press Release NO. 64/37, dated October 2, 1964, p.1.

[12] Central African Power Corporation, 2nd Annual Report and Accounts for the year ended 30th June 1965 p. 17 and p. 30.

[13] G. Schwarzenberger, *International Law as Applied by International Courts and Tribunals,* Vol. I, 1959, p. 37.

[14] H. A. Smith, *The Economic Uses of International Rivers,* 1931, pp. 1–4.

[15] *Op cit.* in note 12 above, p. 218.

[16] Report to the Power Committee of the Economic Commission for Europe, on '*Legal Aspects of Hydro-Electric Development of Rivers and Lakes of Common Interest*'. ECE Doc. E/ECE/136 of 1952, p. 1.

[17] Schwarzenberger, *op cit.,* p. 220 et seq.

[18] F. J. Berber, *Rivers in International Law,* 1959, p. 9. The view expressed by the learned author appears to have been confirmed by the events leading up to the conclusion of the Indus Waters Treaty of 19th September 1961. Joint regulation of the Indus waters was vital to the existence of both India and Pakistan, yet only after years of difficult negotiation and the significant intervention of the World Bank was a settlement achieved.

[19] See note 2 above.

[20] See agreement cited in note 6 above.

[21] See Appendix I of the Agreement, containing draft provisions for the dissolution Order.

[22] See Guarantee Agreements (Loan 145 RN) Art. III 3.06.

[23] See Lord McNair, *The Law of Treaties,* 1961, p. 115 where the learned author observes 'It is not at the moment easy to give a simple answer to the question whether agreements between the Commonwealth countries are governed by international law or some domestic system of law. In each case the answer must depend upon the intention of the parties. There is no reason why they should not enter into agreements which they regard as not being within the field of international law, and there is no doubt that they are free to contract, and capable of contracting, obligations with one another based on, and governed by, international law.'

[24] Parliamentary Debates (Hansard) House of Commons Official Report, 1/12/1965, col. 1438–9.

[25] Art. III para 3.06. of the Agreements between the Bank and Southern Rhodesia and the Bank and Northern Rhodesia. Also to be found in the 1964 Guarantee Agreements between the Bank and the two governments. Loan No. 392. RNS.

[26] Sec. 43. Dissolution Order of 1963.

[27] Central African Power Corp, 2nd Annual Report, 1965, p. 1.

[28] *Ibid.*, p. 20.

[29] *Ibid.*, p. 15.

[30] Art. 2. Intergovernmental Agreement cited in note 6 above.

[31] Sec. 34 (2) of National implementing laws cited in note 7 above.

[32] Art. 4 (a) November Agreement cited in note 6 above.

[33] Art. 4 (b) *ibid.*

[34] Art. 4 (b) of November Agreement cited in note 6 above.

[35] Sec. 34(2) Dissolution Order-in-Council, Art. 20 November Agreement.

[36] Art. 20 (b) November Agreement.

[37] Art. 20 (e) *ibid.*

[38] See G. Schwarzenberger, *The Inductive Approach to International Law,* 1965, pp. 100 et seq.

[39] Art. III 3.02 (c) of Loan Agreement and Loan Assumption Agreements relating to Loan 149 RN, between the Bank and the Federation and between the Bank and the two Territories. Also the identical article in the 1964 Loan Agreement (392 RNS). See also Bank Press Release No. 65/64 of December 14, 1965.

[40] Art. V. 5.08 Bank—Central African Power Corporation Agreement. No. 392 RNS.

[41] Loan Regulations No. 4 of the Bank dated February 15, 1961, as amended by Schedule 3 of Loan No. 392 RNS. See also Broches, 'International Legal Aspects of the Operations of the World Bank' 98 *Hague Recueil,* 1959, p. 297.

[42] Art. 74. Charter of the United Nations, 'Members of the United Nations also agree that their policy in respect of the territories to which this Charter applies, no less than in respect of their metropolitan areas, must be based on the general principle of good-neighbourliness, due account being taken of the interests and well-being of the rest of the world, in social, economic, and commercial matters.'

[43] But see Schwarzenberger *op cit.* at p. 218.

CHAPTER X

# Outline Programme for Hydro-Electric Development in West Africa to 1980

STANTON R. SMITH[1]

## INTRODUCTION

This study presents a preliminary outline of a long-range development programme for West African electric generation and transmission systems, making the optimum use of the region's rich water resources. The preparation of long range perspective development plans makes possible the assessment of overall electric energy investment requirements over the next fifteen years and the evaluation for consistency and coordination of individual projects within the context of overall requirements. This study attempts only a first approximation of a perspective plan based upon available load growth and engineering information and will require continuous review and updating as additional information becomes available from the individual West African countries' hydro-electric development programmes.

## WEST AFRICAN HYDRO-ELECTRIC POTENTIAL

The basic and most economic source of electric energy in West Africa for the foreseeable future is the development of the vast undeveloped hydro-electric potential of its major rivers, including the Niger, Benue, Volta, Senegal, Bandama, Cavally, Konkoure, Sassandra and Comoe.

Africa is rich in hydro-electric potential—the entire continent has one-third of the world's prime hydro-electric potential capability, 176,677 MW[2] of the total 551,605 MW (reference Table I). However, Africa has only one-fiftieth of the world's total installed capacity.

Unlike Europe and North America the hydro-electric potential of West Africa has only begun to be developed in the past decade. It may be noted from Table 1 that 41 per cent of the total hydro-electric potential has been developed in Europe and 23 per cent in

North America—the figures are much higher for Western Europe and the United States where there are very few economically developable sites left. However, in Africa as a whole and in West Africa in particular, of the total hydro-electric potential, only a few of the prime sites have been utilised; there are hundreds of economic low-unit-cost sites awaiting development.

In West Africa the flow of two of the largest rivers, the Niger and Volta, has been completely controlled by two large dams, the 768-MW Akosombo and the 960-MW Kainji. There are also existing smaller hydro-electric developments at Ayamé I and II (50 MW) in the Ivory Coast and Mt. Coffee (34 MW) on the St. Paul River in Liberia. However, the total installed hydro-electric capacity in West Africa today, including the initial 320-MW installation at Kainji is only 973-MW or only 1 per cent of the region's vast potential totalling 98,440 MW. (Reference Table 1). As a consequence there are many highly economic sites in West Africa for hydro-electric development which have been the subject of varying degrees of engineering study. The new generating capacity of the 15 projects listed below totals over 5,000 MW which, as will be presented later, is sufficient to supply economically, the entire electric energy needs of West Africa until 1980 in a staged development programme including the provision of adequate national and regional transmission grid systems. These sites include:

1. In *Nigeria*: the Niger River Kainji Dam first phase of 320 MW will be expanded to 960 MW. There are many low-cost additional sites including two located on the national 330 KV transmission grid, which have been recommended by the Engineer for installation, the 480-MW site at Jebba on the Niger below Kainji, and a 500-MW site on the Kaduna River at Shiroro Gorge.
2. On the Mono River bordering *Togo* and *Dahomey* there are four sites with a total capability of 115 MW, the subject of a recently completed U.N. engineering study recommending its construction and the Ouemé River in Dahomey has 700 MW potential.
3. In *Ghana* the initial installation of 512 MW at Akosombo Dam will be increased to 768 MW with two more 128-MW units; below Akosombo at Kpong there is a low-cost site for a weir with 86–MW capability and the USSR has designed a dam of approximately 120 MW at Bui on the Black Volta towards the *Ivory Coast* border.

4. The 174-MW Kossou Dam at Kossou on the Bandama River studied by Kaiser Engineers and Electricité de France for the *Ivory Coast* government. Other sites which have been preliminarily investigated by Electricité de France or Kaiser Engineers, include a 360-MW site at Tiboto on the Cavally River bordering *Liberia,* a 120-MW site at Attakro and a 240-MW site at Malamalasso on the Comoé River, a 150-MW site at Brimbo on the Lower Bandama River, apart from other sites on the Sassandra River.
5. In *Liberia,* the 34-MW Mt. Coffee can be increased to 102-MW with upstream storage and the Cavally River, Tiboto Project mentioned above could follow.
6. The 1,200-MW Konkouré River Project in *Guinea* including the 480-MW Souapiti dam and the 720-MW Amaria dam, has since the 1950s been proposed as the source of power for a major aluminium industry based upon rich domestic bauxite deposits.
7. The 300-MW Gouina Dam on the Senegal River in *Mali* is the keystone of a major multi-purpose river development programme presently under study with UN financing; other planned hydro-electric projects to follow on the Senegal River total 580-MW, including the 100-MW Bakel site between *Mauritania* and *Senegal.*

RECENT HYDRO-ELECTRIC EXPERIENCE IN AFRICA

In the 1950s several of Africa's major hydro-electric schemes which offered the possibilities of low-cost electricity of less than 4 mills (tenths of U.S. cents) per KWh for large industrial plants consuming at high load factors, were viewed in terms of developing an electro-metallurgical industry, usually aluminium, of sufficient size to justify the basic hydro-electric development. Thus, in 1960, a number of hydro-electric projects were being considered for financing, promoted by joint ventures of the handful of major international aluminium companies including the 1,200-MW Konkouré Project in Guinea, the 768-MW Volta River Project Akosombo Dam in Ghana, the 300-MW Kouilou Project in Congo (Br.), the fabulous 30,000-MW Inga Project in the Congo and a project on the controlled Lower Zambesi River in Mozambique.

Other African hydro-electric projects, including the Kariba Project had been developed on the basis of the existing electro-metallurgical demand of the Zambian copper industry, even in the face of economical thermo-electric generation utilising low-cost Wankie coal. The large 763-MW Congo (DR) hydro-electric developments also fed electro-metallurgical industry. Uganda's Owen Falls hydro-electric project on the Upper Nile at the mouth of Lake Victoria proceeded on the basis of optimistic anticipations of electro-metallurgical industrial development, and was subsequently economically justified only by construction of a transmission link to supply the Nairobi demand in Kenya. The highly-economic Aswan Dam Project in Egypt founded its justification on supplying low-cost electricity to an existing large-scale electric generating system, including projected electro-metallurgical industry development, but more importantly on the complete control of the waters of the Lower Nile to allow an effective 50 per cent expansion of the total annual multi-crop irrigable acreage in Egypt, which must feed a population of 29 million.

Needless to say, developments since 1960 have not borne out Africa's optimistic hopes for major hydro-electric development. Of the half dozen major projects dependent upon the development of new electro-metallurgical industrial complexes, only the Volta River Project in Ghana has been brought to life, and this occurred only after overcoming political opposition and in spite of the fact that the initial smelter size was reduced following the withdrawal of support by the largest of the proposed aluminium joint venture partners.

The only other major African hydro-electric project internationally financed after Akosombo was Nigeria's Kainji Project, reflecting the large size of Nigeria.

### LARGE SIZE OF AFRICAN HYDRO-ELECTRIC PROJECTS

The optimum (lowest cost per KW) hydro-electric projects in Africa at this early stage of development tend to be large-scale, reflecting the larger size of Africa's rivers and the fact that initial projects on each river must be planned and constructed for full river control. This has meant that the economic justification of those hydro-electric projects which have proceeded in Africa in the last five years has been largely dependent upon the construction of transmission grid systems unifying larger potential markets. The following five examples will illustrate this point:

1. In Ghana, while the initial 115,000 ton/annum aluminium smelter constructed by Kaiser at Tema with its 200-MW initial demand was critical to the initiation of the Volta River Project, the construction of a 510-mile national transmission grid system unifying another 100 MW of domestic

demand throughout Southern Ghana was necessary to make the first stage 512-MW installation at Akosombo economic;

2. the 600-MW installation at Kariba was dependent upon an international transmission grid system linking more than 200 MW of high-load-factor electro-metallurgical electricity demand from the Zambian copper mines with the urban demand of Lusaka and of the major industrial centres of Rhodesia, including Bulawayo and Salisbury. The subsequent placing of the existing coal-fired thermal generating plants in Rhodesia and Zambia on stand-by has resulted in the full utilisation of Kariba's power plant next year, five years after construction;
3. the 120-MW Owen Falls hydro-electric initial installation was only partially utilised by demand from the Uganda national transmission grid system as over optimistic views of potential industrial demand were followed only by the construction of a smaller-scale electric steel mill and copper smelter. As a result, two years after the construction of the dam, an international transmission link with Nairobi was constructed and today within six years of initial operation the last two 15-MW units are being installed to complete the installation's 180-MW design capacity. However, lower cost hydro developments on Uganda's Upper Nile are now being bypassed for Kenya's Tana River hydro-electric development;
4. in Nigeria, a 1,200 mile national transmission grid system linking all major cities of the country including Lagos, Ibadan, Enugu, Kano, Kaduna and Port Harcourt is being constructed to provide sufficient demand to justify the initial installation of 320 MW at Kainji;
5. For the proposed 174-MW hydro-electric station at Kossou on the Bandama in the Ivory Coast, a $12 million national transmission grid system over 200 miles long is proposed linking all consuming centres from Bouaké to Abidjan to Ayamé.

### ADVANTAGES OF TRANSMISSION GRID DEVELOPMENT

Transmission grid systems inter-tying larger numbers of small consuming and producing centres offer a number of economic advantages for more economical power supply and make possible the construction of large-scale, lower cost hydro-electric developments:

1. They bring together larger aggregations of demand, making possible the installation of larger and more efficient generating stations; this is especially valuable in Africa where loads tend to be low and smaller centres with demands less than 15 MW are forced to rely on high-cost small-scale diesel electric generation stations rather than benefiting from the economies of larger-scale steam generators or hydro-electric developments;
2. the unified transmission grid system minimises the need for reserve and stand-by generating capacity. Small individual stations must each maintain stand-by generating capacity equivalent in size to the largest single generating unit, while one such stand-by generating unit may serve an entire grid system, reducing the total requirement for generating capacity in the country for any given level of demand;
3. aggregating electricity demand centres means that higher system load factors are possible as a result of combining industrial and the air conditioning loads with domestic lighting loads and as a result power generating capability is more efficiently utilised throughout the system;
4. dependability of supply is increased in all consuming centres removing the necessity for individual industrial installations to install their own expensive stand-by generating capability;
5. more favourable system load factors and generator utilisation makes possible lower general tariff structures, lower cost for new connections, and more importantly, lower tariffs for industrial users with higher load factors, thus encouraging industrial development;
6. the construction of new generating stations can be undertaken at their optimum location, including both tide-water locations for steam electric generating stations depending upon imported fuel, and hydro-electric sites developed without unduly penalising the individual project's economics by loading its costs with the whole transmission investment;
7. optimum use can be made of existing generating stations by centralised control, concentrating production in the most efficient stations, and increasing the firm-power capability of individual hydro-electric stations by operation in conjunction with thermo-electric stations;
8. man-power requirements and operating costs for the entire production and distribution systems are minimised as the

duplication of smaller isolated generating and distribution centres is removed.

SOME ECONOMIC DISADVANTAGES OF HYDRO-ELECTRIC DEVELOPMENTS

There are a number of factors which should be considered in the economic evaluation of the use of Africa's hydro-electric resources in a long-run electric power development programme. Some of these factors are:

1. The size of individual hydro-electric projects tends to be large both in comparison with total demand in individual African countries, and in comparison with the annual increase in electric capacity requirements. On the demand side, African electric systems have been growing rapidly in recent years and transmission grid systems have been built aggregating larger blocks of demand. However, on the supply side, the disability of large initial size of installation, and therefore investment, cannot be easily overcome since the size of the dam and spillway structure are determined by the characteristics of the river flow, including the maximum flood size. The size of the initial dam is also governed by the desirability of gaining complete control of the annual flow cycle of the river so as not to prejudice the future full use of the river's hydro potential. After the first dam is constructed, the cost per KW of subsequent dams on the lower river is significantly less because the river control achieved by the initial dam allows maximum continuous electricity production with a minimum reservoir size and often a smaller spillway provision. This future benefit, however, is seldom reflected in the initial project's economic evaluation. The large size of recent African hydro-electric developments is indicated by comparing Ghana or Nigeria's present 100-MW demand to the capacity of the following hydro-electric projects: Akosombo, 768 MW; Aswan 2,100 MW; Bin El Ouidane 212 MW; Kariba 600 MW (future 1,500 MW); Owen Falls 180 MW; Kainji 960 MW; Kossou 174 MW; Inga 30,000 M.W.
2. The initial capital cost per KW of installed capacity for hydro-electric developments is often higher than that of thermo-electric generating stations, especially if the latter are constructed cheaply without regard to maximising fuel economy, including provision for discharge heat recycling, etc. Thermal all-in capital costs per KW average around

$220 for 30-MW units, including engineering and interest during construction.

The cost of the dam and spillway structures make up the largest proportion of a hydro-electric project's cost and it is this cost which varies significantly with site conditions, while the cost of the basic power generating station equipment on hydro-electric projects varies only moderately, averaging about $50 per KW. An engineering design commentator observed[3] that construction costs of African hydro-electric projects have dropped to the same levels as costs in developed countries. Tables 2 and 3 providing hydro-electric project costs in Africa and throughout the world in recent years indicate that African costs are low considering price trends. The construction cost per KW of African hydro-electric projects varies, but tends to decrease with greater size, as indicated in the following list: the cost based upon ultimate capacity at Aswan is $176 per KW; at Akosombo $195/KW; at Kariba $260/KW; at Kainji $224/KW and at Kossou, Bandama (Projected) $272/KW. Two smaller projects, the 20-MW Ayamé I and 34-MW Mt. Coffee costs were significantly higher at $600/KW and $676/KW, respectively. (The above costs are construction costs only and exclude population relocation and housing, construction community and transmission, a portion of which is generally charged against power).

Given the comparatively low costs of African hydro-electric projects, if capital is available at reasonable interest rates, hydro-electric power should be cheaper in Africa than thermo-electric generation, when the disability of under-utilisation in the early years can be minimised by industrial developments or by aggregating larger loads with transmission systems.

3. The location of the hydro-electric site is determined by topographical factors and may be more or less isolated from the centres of electricity consumption. Isolation of the hydro-electric site at times makes it necessary to install longer transmission systems to reach the centres of consumption. An offsetting factor is that the transmission system often allows the consolidation of load centres into larger aggregates of consumption. The extent to which these transmission systems must be considered a charge against the hydro-electric project is subject to examination.

PROPOSED INTERNATIONAL ELECTRIC TRANSMISSION INTER-TIE SYSTEMS IN WEST AFRICA

From the above background we can now turn to a discussion of specific international transmission grid system developments in West Africa. There are several of these projected for the 15 years ahead, including the linking of the following countries:

1. *Ghana, Togo, and Dahomey* in the immediate future by 161 KV line from Akosombo to Ho-Lome-Cotonou-Porto Novo, and by 1975 to Lagos linking *Nigeria*
2. *Ghana to Ivory Coast* by 1970 via 161 KV line from Prestea to Ayamé
3. *Mali to Senegal, Mauretania and Gambia* by 1975 with 330 KV line from the Gouina hydro-electric station in Mali to Dakar, Bamako, Bathhurst and Nouakchott
4. *Ivory Coast to Liberia* by 1980 with 330 KV link from border Cavally River Hydro-electric Station to Buchanan and Abidjan
5. *Ghana-Upper Volta* link by 1980 from Bui Hydro-electric Station on Black Volta River with 161 KV transmission link via Tamale and Bolgatanga to Ouagadougou.

In Ghana, the Akosombo Dam of the Volta River Project has been completed with a 512-MW first-stage capacity, and is presently serving by way of a national transmission grid system the southern half of Ghana, including the communities of Tema, Accra, Takoradi and Kumasi. The first stage development of Akosombo is sufficient to carry the Ghana electricity demand until 1969, as may be seen in the following table:

| | |
|---|---|
| Existing Installation (4 × 128 MW) . . | 512 MW |
| Initial Smelter Demand (115,000 tons) (early 1967) . . | 200 MW |
| Committed to 2nd Stage Smelter (1969) | 100 MW |
| Existing Ghana Grid Demand (incl. 10-MW thermal supply) . | 71 MW |
| 1969 Additional requirement for Ghana Grid . | 22 MW |
| Reserve Requirement . . | 128 MW |
| | 521 MW |

While the initial capability of the Akosombo project will be utilised in the next three years, there is a capacity in the power station for two more 128-MW generators, and this additional power will be entirely available because there will be no further need for additional reserve of capacity on the Ghana interconnected system. Therefore, the installation of a single additional generating unit will give 128 MW of additional available supply. However, because of the economies of equipment and installation cost, both additional units would probably be installed at the same time.

Ghana's two neighbouring countries, Togo and Dahomey, lying nearest the Akosombo dam could be served by a transmission link. Akosombo power is fed to the national transmission grid system at 161 KV, and it would be possible to add immediately an additional bay at Akosombo to take off power at 161 KV to serve a long-distance transmission line via Ho in Ghana, which must be served in any event, to Lome in Togo, a total distance of 90 miles, and then another 100 miles to serve Cotonou and Porto Novo in Dahomey.

We should note, however, that the 1966 demand in Togo and Dahomey is low, approximately 35 million KWh in Togo and 20 million KWh in Dahomey, and that the economic justification for this transmission link with the Akosombo project would have to be in terms of delivering energy at below the approximately 35 mills per KWh cost of diesel electric generation in the existing load centres.

An engineering study is necessary to provide accurate costs for an Akosombo-Ho-Lome-Cotonou-Porto Novo transmission link; however, the table below indicates the order of magnitude of the capital cost of the project and annual economic charges.

*Capital Costs*

| | |
|---|---|
| 190 miles of 161 KV line . . | $4,200,000 |
| 3 sub-stations . . . . . | 1,100,000 |
| Sub-total | $5,300,000 |
| Plus 30% . . . . . . (Engr., Contingency and Interest) | 1,600,000 |
| Total Transmission Capital Costs | $6,900,000 |

*Annual Costs*

| | |
|---|---|
| Annual Capital Costs . . . | $ 480,000 |
| (6% & 35-year life ·069) | |
| Annual Operating Costs . . | 165,000 |
| (2·4% × capital) | |
| Total Annual Costs | $ 645,000 |

This high tension transmission link serving Togo and Dahomey could be constructed within a year, and given an existing demand of approximately 55 million KWh we can anticipate a load of approximately 118 million KWh by 1970. Given an annual transmission cost of $645,000, we can place an economic charge on transmission at approximately 8 mills per KWh. The price of Volta energy from the Volta River Project to be added to this transmission cost would have to be the subject of negotiation; however, we might take a figure of 7 mills per KWh for the purposes of this economic exercise, which would give a total average cost of energy in Togo and Dahomey from the transmission grid system of approximately 15 mills per KWh.

The 15 mills per KWh cost for energy in Togo and Dahomey would certainly be attractive compared to the existing cost of diesel electric generation in those countries, and in comparison to the existing tariff structure, which ranges to 100 mills per KWh for the private power company in Dahomey and somewhat less in the public power system in Togo as a result of tariff reductions after nationalisation.

The proposed international transmission grid system linking Ghana, Togo and Dahomey described above will follow the same route as the existing transmission links in those countries and the proposed transmission links which will have to be constructed in any event when larger and more efficient hydro-electric or thermo-electric generating stations are developed in Togo and Dahomey.

At the present time an engineering study for the development of hydro-electric stations on the Mono River serving both Togo and Dahomey is being completed. As part of this programme a high-tension transmission grid system linking the major consuming centres in Southern Dahomey and Togo is proposed on the same alignment as a transmission link to Ghana. In Ghana it is proposed to link the Volta Region, including the city of Ho, with the Akosombo Dam in the near future in any event. Thus, the only additional mileage involved is the link between Ho and the nearest point on the Togo transmission grid system, approximately 20 miles.

The immediate construction of a transmission grid system linking Akosombo with Togo and Dahomey could be viewed as an interim measure to handle the demand in these countries over the next five years instead of installing additional thermo-electric generating capacity until the first stage 31-MW hydro-electric station can be built on the Mono River at Nangebeto. At that point the new hydro-electric development on the Mono River can feed directly into the by then existing international transmission grid system, and can be planned to feed in at the same 161 KV tension.

It may be noted that in two more years the Nigeria national transmission grid system will be completed and connected to the 320 MW first-phase Kainji hydro-electric station. At this time, in order to increase dependability and stability, the Ghana-Togo-Dahomey transmission grid could be connected to the Nigerian system across from Porto Novo to Lagos, a distance of 50 miles. This international inter-tie would not only assure stability, but at the same time allow for future interchange of blocks of power between the Nigerian and Ghanaian electrical systems.

In Ghana, it is anticipated that the entire 768-MW capability of the Akosombo dam will be utilised in the mid-1970s. After that the Kpong Weir can add another 86 MW, and Bui hydro-electric project 119 MW. However, by the end of the 1970s, the entire hydro-electric capability of the Volta River, totalling approximately 973 MW, will be utilised. This anticipates that the aluminium electro-metallurgical complex centred in Tema will expand towards 330,000 tons per year, requiring 600 MW, and that Ghana's bauxite deposits would be exploited to serve this aluminium complex. At Kibi, near Korforidua, two hills of bauxite have recently been geologically investigated preparatory to mining development, to support an alumina industry. The proven reserves on these two hills total 80 million tons of high-grade (45 per cent alumina) bauxite capable of supporting an expanded aluminium industry at Tema for 50 years (4 tons bauxite = 1 ton aluminium). The development of the alumina industry will add another 30 to 40 MW of electricity demand to the Ghana system.

Given the above projections of demand for electricity in Ghana, it is apparent that the entire hydro-electric resources of the Volta River will be utilised by 1980. It is therefore anticipated that after that period it will be the vast low-cost hydro-electric capability of the Niger River which will be supplying large blocks of power to the electro-metallurgical industrial complex at Tema over an international extra-high-voltage long-distance transmission link. We have mentioned earlier some of the obvious immediate stages of hydro-electric development which follow the first 320-MW installation in Kainji including the 640-MW second stage, 480 MW at Jebba, and 500 MW at Shiroro Gorge.

### WEST AFRICA ELECTRIC SYSTEM DEVELOPMENT PROGRAMME TO 1980

The requirements for electric energy in West Africa over the next fifteen years would be most economically supplied by development of the hydro-electric sites listed on pp.159–160. It may be noted that the cost of construction of the hydro-electric developments above tends to be low in comparison with hydro-electric projects constructed throughout the world during the past decade. This may be seen by examination of Table 2, which provides a list of recent international hydro-electric projects and their cost of construction per KW, along with a list of the costs of recent African hydro-electric projects. The only significant alternative electricity supply source is higher cost steam thermo-electric generation, which in West Africa, outside gas-rich Nigeria, must be fired with imported oil.

While the initial capital costs of thermo-electric stations are often somewhat lower than even the favourable hydro-electric projects listed above, an economic comparison of generating costs utilising reasonable long-term rates of interest indicates that the long-run savings in fuel and operating costs of the hydro-electric developments provide lower average electricity costs. Financial sources in the industrialised countries, however, have recently tended to place more stringent economic requirements on hydro-electric projects than on thermo-electric projects with the stated reason of conserving capital. The necessarily arbitrary requirements include the understatement of the value of other river control benefits, and the requirements that these projects should be able to bear an interest rate higher than the normal 6 per cent. Some possible reasons for this hyper-critical evaluation of hydro-electric developments are discussed in the last section of this essay but for the programming purposes of this study, West Africa's power requirements are assumed to be supplied by the lowest cost generating source, without regard to its capital intensity.

The first requirement for programming future electric generating requirements is the preparation of projections of electricity demand for future years. Table 4 presents the trends in electric power consumption over recent years, and projections of requirements for the years 1970, 1975 and 1980 in each of the West African countries.

The electric load growth projections have been prepared utilising conservative annual growth rates of 10 to 15 per cent per annum, which are considerably below the rates of growth experienced in West African countries in recent years. Electricity load growth for the entire region has averaged 17.5 per cent per annum since 1959, in spite of high connecting charges and domestic electric tariff rates ranging to 10 U.S. cents per KWh in most of the francophone countries. Every West African country experienced electricity consumption growth rates above the world average of 7 to 8 per cent between 1959 and 1965, as may be seen in the following list of growth rates:

| | | | |
|---|---|---|---|
| Nigeria | 22·0% | Ivory Coast | 27·1% |
| Dahomey | 20·0% | Liberia | 22·9% |
| Togo | 36·8% | Sierra Leone | 15·4% |
| Ghana | 9·2% | Guinea | 35·6% |
| Upper Volta | 20·3% | Gambia | 7·9% |
| Niger | 18·2% | Senegal | 9·2% |
| Mali | 9·8% | Mauritania | 84·6% |

While electricity consumption grows at approximately the world average annual rate in the United States, which produces half the world's supply, the rates of growth in smaller developing countries' systems are typically much higher. There is an inverse relationship between electricity consumption growth rates and per capita income—the lower the per capita income the higher are electricity growth rates reflecting the fact that per capita electricity consumption is increasing rapidly from very low levels. The positive correlation between per capita electricity consumption and per capita income for 100 countries is shown in Figure I. West African per capita electricity consumption figures are low ranging from 3 to 8 KWh per annum in Upper Volta, Niger, Mali and Dahomey to 67, 79 and 260 KWh in Ghana, Senegal and Liberia, respectively.

In four countries, Ghana, Ivory Coast, Liberia and Guinea the demand projections include significant industrial power demand requirements. In Ghana the aluminium industry requirement for the Tema smelter now under construction is anticipated to increase from the 1969 requirement of 300 MW to 600 MW by 1980. In Guinea the aluminium industry is anticipated to require 200 MW by 1975 with the construction of a 110,000 tons per annum smelter.

However, no allowance has been made for the important future load to be provided by the electrification of West Africa's railway systems. Electrification is economically advisable on most systems today, not only to utilise cheaper domestic hydro energy instead of imported oil, but to increase West Africa's already pressed single-track railway carrying capacity and delay the capital expense of double tracking.

The requirement for electric power in the West African region is anticipated to quadruple from the total of 1,589 MW installed in 1965 (including the 512 MW at Akosombo) to 6,442 MW in 1980. The demand for electric energy will increase faster from the 3,038 GWh (millions of KWh) consumed in 1965 to 8,608 GWh in 1970 and 30,816 GWh in 1980. Installed generating capacity increases less rapidly than energy consumption because of more efficient use of generating capacity, including the high rates of production of the anticipated 800 MW feeding Ghana's and Guinea's aluminium industries at 95 per cent load factor in 1980.

To utilise optimally West Africa's hydro-electric resources, the high voltage electric transmission system will have to be expanded and intertied increasing the total length from the 4,272 miles in 1965 to 8,948 miles in 1980.

On the basis of the projected power requirements for each West African country an optimum electric generating programme, including transmission system requirements, has been prepared for each of these countries for the five-year periods 1966-1970, 1971-1975 and 1976-1980. This programme is presented in Table 5.

In the 15-year period to 1980, the West African Region will require:

553 MW of new thermal power costing $104 million,

5,016 MW of new hydro-electric power costing $1,163 million,

3,526 miles of high tension transmission lines costing $124 million,

for a total new investment requirement of $1,391 million.

A map of West Africa has been prepared showing capacities of the existing electric generating stations and transmission systems, and the projected generating station and transmission system requirements for the years 1970, 1975 and 1980.

### INSTITUTIONAL FACTORS HINDERING THE OPTIMUM DEVELOPMENT OF AFRICA'S HYDRO-ELECTRIC RESOURCES

In recent years it would appear that hydro-electric projects in Africa (and other developing countries) have been subjected to more stringent economic evaluation standards by financing sources in the industrialised countries. These standards are typically higher than those applied to hydro-electric projects in the industrialised countries themselves. These more stringent standards take two forms:

1. A tendency to ignore or understate in the developing countries the other benefits of river control, including irrigation and urban water supplies, flood control and reclamation, navigation and recreation.
2. The demand that the hydro-electric project be cheaper than the thermo-electric alternative at interest rates substantially above the actual cost of money, or long-term interest rate of, say 6 per cent. Requirements of 'internal rates of return' for hydro-electric projects have been established as high as 14 per cent by the international leading banks.

An interest return requirement of 14 per cent may be appropriate to allocate or limit loanable funds among the all too prevalent real estate and commodity speculation activities in developing countries, but not to build the basic industry required to transform their economies. Three per cent interest rates were used in the economic justification studies and in the financing of most of America's great hydro-electric projects, including Grand Coulee, Boulder Dam and the TVA, built in the depressed and under-developed regional economies of the 1930s. When they were being constructed three decades ago there were many Cassandras, including professional economists, critical of these projects—today, after they have repaid their original investment a hundredfold in direct and indirect economic benefits, there are no critics left. Today's critics of Africa's hydro developments should respect the lessons of history; Africa must develop her hydro resources at this stage of her economic development to capture the manifold benefits for the next generation. While the lesson of history above was drawn from America's experience, the lesson from Soviet experience is the same. The Soviet Union made heroic sacrifices to harness her water resources from its earliest years and today has an efficient hydro-electric system, which includes the world's largest hydro-electric station at Bratsk with an installed capacity of 6,000 MW.

It is not questioned that all industrial projects in developing countries, including hydro-electric projects, should be subjected to competent economic analysis reflecting the time value of money in the project's capital and operating cost analysis, and in measuring its future benefits or the alternative cost of supply. However, the setting of arbitrary artificially high project evaluation standards for hydro-electric project loans serves to prejudice the developing countries' economic development, including the optimum long-run utilisation of water resources.

Most power technicians in Western developed countries have gained their training only in thermal power technology as a result of constructing and operating thermal installations. The overwhelming proportion of new electric power projects in Western Europe and North America are steam thermo-electric generating stations which have increased in efficiency with increased size over the years. It is understandable that these technicians erroneously favour power development programmes consisting of staged smaller-scale thermo-electric generators as the economic alternative to developing West Africa's hydro-electric potential.

However, the question remains as to why higher arbitrary standards have been set for hydro-electric projects in developing countries, and there are important quasi-political factors inhibiting the development of Africa's hydro-electric projects which must be considered in planning Africa's future power supply programme. There are a number of reasons why financing sources in industrialised countries may tend to favour thermo-electric generating projects over hydro-electric:

1. For thermal projects a larger proportion of the total project cost and, therefore, loan funds tends to be spent on procuring equipment and industrial supplies from supplying industries in the industrialised countries, in contrast to hydro-electric projects where larger proportions are spent on construction activity. The money which will be spent for foreign construction services does not bring with it as much concentrated political support in the industrialised countries because construction firms in these countries are not as politically influential as are equipment suppliers. The major equipment suppliers rank among the top 100 firms in the industrialised countries, whether classified as to capital employed or value added, and wield proportionate political power, whereas not even the largest construction firms rank among the top 100.

2. Each hydro-electric generating station replaces petroleum sales. For example, Ghana would have had to spend up to $35 million per year on fuel oil for steam electric generating stations to produce the total capability of the Volta River Project. Needless to say, such an important and expanding market for petroleum products in African countries is not sacrificed by vested interests without considerable opposition.
3. Hydro-electric projects are usually developed under public rather than private ownership institutional arrangements. Most hydro-electric developments have multi-purpose benefits through river control, including irrigation, flood control, water supply and navigation. In addition, the river control achieved by the initial project often makes possible lower unit cost hydro-electric developments in subsequent projects on the same river system. This means that generally the hydro-electric development is more appropriately the subject of public ownership and control, which engenders elements of conflict from private institutions whether they be domestic private power generation and distribution companies, as exist in some African countries, agricultural plantation and industrial owners, or the western industrial lending countries who are generally concerned about public ownership and operation.

[1] The views expressed in this study are those of the author and do not necessarily represent those of Kaiser Engineers International or The African Development Bank.

[2] Conventional electric engineering abbreviations are utilised in this study. MW stands for a Megawatt, referring to the generating capacity of a thousand KW (Kilowatts); KWh stand for a Kilowatt-hour and GWh for a Gigawatt-hour, or one million Kilowatt hours.

[3] Articles in the engineering journal *Water Power* of May and June, 1965 entitled 'Volta River Project—Akosombo Dam' by Ian Mackintosh, M.A., A.M.I.C.E.

TABLE 1

AFRICA AND WORLD HYDRO-ELECTRIC INSTALLED AND POTENTIAL CAPACITY, 1963

| *Area or Country* | *Installed Capacity MW*[1] | *Estimated* *Hydro-electric Q 95* | *Potential, MW*[2] *Arithmetic Mean* |
|---|---|---|---|
| *America* | *68,095* | *140,815* | *741,359* |
| North and Central | 61,230 | 90,065 | 270,009 |
| South | 6,865 | 50,750 | 471,350 |
| *Africa* | *3,714* | *176,677* | *684,690* |
| West | 653 | 21,400 | 98,440 |
| East | 1,028 | 30,770 | 195,800 |
| Central | 1,065 | 117,300 | 330,000 |
| North | 893 | 1,940 | 29,730 |
| South | 75 | 5,267 | 30,720 |
| *Asia* | 19,992 | 160,826 | 944,153 |
| *Australia and Oceania* | 3,860 | 18,600 | 143,750 |
| *Europe, including U.S.S.R.* | 85,806 | 54,687 | 209,505 |
| Total World | 181,467 | 551,605 | 2,723,457 |
| Africa's Percent of Total World | 2.0% | 32.2% | 25.0% |
| *West African Countries* | *653* | *21,400* | 98,440 |
| Dahomey | 0 | 600 | 2,240 |
| Gambia | 0 | n.a. | n.a. |
| Ghana | 512 | 1,500 | 7,500 |
| Guinea | 21 | 500 | 8,000 |
| Ivory Coast | 50 | 500 | 7,300 |
| Liberia | 38 | 4,250 | 7,500 |
| Mali | 5 | 750 | 4,400 |
| Mauritania | 0 | 200 | 2,500 |
| Niger | 0 | 500 | 12,000 |
| Nigeria | 25 | 9,500 | 22,000 |
| Portuguese Guinea | 0 | n.a. | 150 |
| Senegal | 0 | 500 | 5,500 |
| Sierra Leone | 0 | 2,000 | 3,750 |
| Togo | 2 | 100 | 600 |
| Upper Volta | 0 | 500 | 15,000 |
| *East African Countries* | *1,028* | *30,770* | *195,800* |
| Ethiopia | 9 | 4,250 | 35,000 |
| French Somaliland | 0 | n.a. | n.a. |
| Kenya | 6 | 1,500 | 16,800 |
| Malagasy | 24 | 14,300 | 80,000 |
| Mauritius | 14 | 20 | 100 |
| Reunion | 4 | 20 | 100 |
| Rhodesia and Nyasaland | 810 | 4,680 | 22,000 |
| Somali (Republic Somalia) | 0 | n.a. | 300 |
| Tanganyika | 40 | 3,000 | 26,000 |
| Uganda | 121 | 3,000 | 15,000 |

Table 1—*continued*

| *Area or Country* | *Installed Capacity MW*[1] | *Estimated Hydro-electric Q95* | *Potential, MW*[2] *Arithmetic Mean* |
|---|---|---|---|
| *Central African Countries* | *1,065* | *117,300* | *330,000* |
| Angola | 120 | 4,250 | 78,300 |
| Burundi | 0 | n.a. | n.a. |
| Cameroon Federation | 159 | 4,800 | 28,700 |
| Central African Republic | 4 | 3,500 | 13,800 |
| Congo | 763 | 97,000 | 180,000 |
| Gabon | 19 | 6,000 | 21,900 |
| Ruanda | 0 | n.a. | n.a. |
| Spanish Guinea | 0 | 750 | 3,000 |
| Tchad | 0 | 1,000 | 4,300 |
| *Northern African Countries* | *893* | *1,940* | *29,730* |
| Algeria | 180 | 225 | 6,000 |
| Canary Islands | 1 | n.a. | n.a. |
| Egypt | 365 | 375 | 900 |
| Libya | 0 | n.a. | 200 |
| Morocco | 320 | 300 | 1,500 |
| Spanish West Africa | 0 | 260 | 750 |
| Sudan | 0 | 750 | 20,000 |
| Tunisia | 27 | 30 | 380 |
| *Southern African Countries* | *75* | *5,267* | *30,720* |
| Basutoland | 0 | 310 | 490 |
| Bechuanaland | 0 | 22 | 3,730 |
| Mozambique | 70 | 3,750 | 15,000 |
| South Africa | 5 | 335 | 10,000 |
| South West Africa | 0 | 150 | 1,500 |
| Swaziland | 0 | 700 | n.a. |
| Total Africa | 3,714 | 176,677 | 684,690 |

Notes:

1. In production or near completion as of early 1963.
2. Total theoretical potential expressed in MW of generating capability available 95 per cent of the time and as the arithmetic mean.

Source:

Adapted from Lloyd L. Young, *Summary of Developed and Potential Waterpower of the United States and Other Countries of the World, 1955–1962*. Geological Circular 483. U.S. Geological Survey, Washington, 1964, pp. 15–16.

TABLE 2

CONSTRUCTION COSTS OF HYDRO-ELECTRIC PROJECTS

| *Year of of Construction* | *Name* | *Location* | *Capacity Capacity MW* | *Construction Cost Millions of $* | *Cost per unit, $/KW* |
|---|---|---|---|---|---|
| 1951 | Davis | Arizona | 225 | 63 | 281 |
| 1952 | Hungry Horse | Montana | 285 | 80 | 281 |
| 1952 | Bull Shoals | Arkansas | 162 | 49 | 301 |
| 1953 | Canyon Ferry | Montana | 50 | 21 | 420 |
| 1953 | Yale | Washington | 108 | 36 | 330 |
| 1953 | Clark Hill | South Carolina | 282 | 75 | 267 |
| 1952 | Cabinet Gorge | Montana | 200 | 45 | 223 |
| 1954 | Lookout Point | Oregon | 120 | 42 | 311 |
| 1955 | Montgomery Creek | California | 90 | 26 | 290 |
| 1955 | Roanoke Rapids | North Carolina | 100 | 31 | 310 |
| 1956 | Palisades | Idaho | 114 | 15 | 135 |
| 1956 | Gavins Pt | South Dakota | 100 | 35 | 354 |
| 1957 | Old Hickory | Tennessee | 100 | 33 | 328 |
| 1958 | Balch 2 | California | 97 | 11 | 110 |
| 1958 | Caribou 2 | California | 110 | 28 | 260 |
| 1958 | Haas | California | 135 | 27 | 197 |
| 1958 | Poe | California | 124 | 37 | 300 |
| 1958 | Brownlee | Idaho | 360 | 69 | 190 |
| 1958 | Pelton | Oregon | 108 | 20 | 188 |
| 1959 | Noxon | Montana | 283 | 84 | 296 |
| 1959 | Upper Baker | Washington | 94 | 46 | 451 |
| 1960 | Fremont Camp | California | 48 | 23 | 496 |
| 1961 | Oxbow | Idaho | 190 | 47 | 250 |
| 1961 | Yellowtail | Montana | 200 | 40 | 249* |
| 1961 | Binga | Philippines | 111 | 51 | 462 |
| 1960 | Serre-Poncon | France | 324 | 102 | 314 |
| 1962 | Barkley | Tennessee | 144 | 27 | 237* |
| 1963 | American River | California | 204 | 92 | 449 |
| 1963 | Green Peter | Oregon | 90 | 25 | 326* |
| 1963 | Guri | Venezuela | 351 | 74 | 260* |
| 1964 | Toledo Bend | Texas | 80 | 22 | 326* |
| 1962 | Monteynard | France | 324 | 49 | 150 |
| 1963 | Angat | Philippines | 222 | 63 | 332* |
| *African Hydro-Electric Projects* | | | | | |
| 1959 | Kariba | Rhodesia | 600 | 240 | 400*** |
| 1959 | Ayame I | Ivory Coast | 20 | 12 | 600 |
| 1965 | Akosombo | Ghana | 768 | 150 | 195** |
| 1966 | Mt. Coffee | Liberia | 34 | 27 | 676 |
| 1968 | Aswan | Egypt | 2,100 | 370 | 176** |
| 1968 | Kainji | Nigeria | 960 | 215 | 224** |
| Projected | | | | | |
| | Kossou | Ivory Coast | 180 | 49 | 272 |

NOTES:

These figures are construction costs per KW and exclude engineering, owner's interest, housing, reservoir costs, transmission, etc. U.S. data are hydropower costs per kilowatt of installed nameplate capacity obtained from the Federal Power Commission's 1957 and 1960 Indexes of Hydro Plant Construction Cost, defined as 'total cost of project, including general overhead construction costs'. Data are not on a strictly comparable basis and indicate order of magnitude only.

*Unit cost is construction bid plus $50 per KW for turbines and generators.

**Estimated cost at ultimate capacity.

***Second stage 900-MW installation would reduce unit cost by a third.

TABLE 3

CHARACTERISTICS OF HYDRO-ELECTRIC POWER PROJECTS IN AFRICA

| *Item* | *Akosombo Ghana* | *Aswan Egypt* | *Ayamé* I *Ivory Coast* | *Bin El Ouidane Morocco* | *Kariba Rhodesia* | *Kossou Ivory Coast* |
|---|---|---|---|---|---|---|
| 1. Installed capacity (kw) | 768,000 | 2,100,000 | 20,000 | 120,000/92,000[4] | 600,000 | 180,000 |
| 2. Annual generation (millions kwh) | 5,400 | 10,000 | 100 | 160/390[4] | 4,000 | 551 |
| 3. Operating hours/year (plant factor) | 7,040 (80%) | 4,760 (54%) | 5,210 (58%) | 1,335/4,240 (15/47%) | 5,720 (65%) | 3,066 (35%) |
| 4. Total Cost (millions U.S. $)[1] | 150 | 370 | 12 | | 240 | 50[5] |
| 5. Energy cost (U.S. $ per kwh)[2] | ·0028 | ·0037 | ·012 | | ·0060 | ·012 |
| 6. Drainage area ($km^2$) | 394,000 | 2,400,000 | 9,320 | 6,400 | 520,000 | 32,400 |
| 7. Storage capacity (millions $m^3$) | 148,000 | 157,000 | 1,075 | 1,160 | 170,000 | 25,000 |
| 8. Reservoir area ($km^2$) | 8,500 | 5,800 | | 33·8 | 4,500 | 1,500 |
| 9. Max. reservoir level (meters) | 84 | 182 | 90 | 810 | 485 | 210 |
| 10. Min. reservoir level ( ,, ) | 76 | 147 | 83 | 765 | 465 | 198 |
| 11. Average head ( ,, ) | 66 | 57 | 19 | 88/232[4] | 87 | 43 |
| 12. Max. flood recorded ($m^3$/sec.) | 14,700 | 16,200 | 700 | 475 | | 1,320 |
| 13. Max. flood calculated ( ,, ) | 34,000 | | 1,200 | 2,500 | | 2,300 |
| 14. Annual flow (millions $m^3$) | 37,000 | 84,000 | 2,580 | 1,090 | | 5,400 |
| 15. Turbines (number, type) | 6 Francis | 12 Francis | 2 Francis | 3/2 Francis[4] | 6 Francis | 3 Kaplan |
| 16. Volume of dam (millions $m^3$) | 8 | 42 | 0·05/0·1[3] | 0·29 | 1 | 4 |
| 17. Type of construction | rockfill | rockfill | buttress/earth | conc. arch | conc. arch | rockfill |
| 18. Crest18. Crest elev. (meters) | 88 | 196 | 93 | 2 | 490 | 212 |
| 19. Crest length ( ,, ) | 640 | 8,000 | 250/380[3] | 312 | 620 | 1,500 |
| 20. Dam height ( ,, ) | 113 | 111 | 29/15 | 133 | 124 | 57 |
| 21. Cost/storage capacity (U.S. $/$10^3m^3$) | 1·01 | 2·36 | 9·30 | | 1·41 | 2·76 |
| 22. Cost/installed capacity (U.S. $/kw) | 195 | 176 | 600 | | 400 | 384[5] |
| 23. Dam volume/storage capacity $\times 10^4$ | 0·54 | 2·71 | 1·56 | 82·5 | ·06 | ·16 |

[1] Including cost of reservoir area (land compensation). [2] Annual cost est. at 10 per cent capital cost.
[3] First figure for concrete part, second figure for wing dams. [4] First figure for power station Bin El Ouidane, second figure for second stage Afourer.
[5] Excluding interest, escalation, reservoir costs, engineering, and certain housing costs: total cost is $49 million and cost per KW is $273.

TABLE 4

ELECTRIC SYSTEM CAPACITY AND PRODUCTION
IN WEST AFRICAN COUNTRIES,
1959 AND 1965 AND PROJECTIONS TO 1980

| *Country and Energy Reserves* | *Year* | *Installations, Mw Hydro* | *Thermal* | *Total* | *Production, Gwh Hydro* | *Thermal* | *Total* | *Per Capita Kwh* | *High Tension (11 KV+) Transmission Length, miles* |
|---|---|---|---|---|---|---|---|---|---|
| *Nigeria* | 1959 | 16 | 136 | 152 | 47 | 318 | 365 | 10·6 | |
| Energy Reserves | 1965 | 28 | 302 | 330 | 125 | 1,081 | 1,206 | 21·4 | 1,460 |
| Hydro—17,000 Gwh | 1959–65 Annual Increase | | | | | .. | 22·0% | | |
| Oil—315 $10^6$ tons | *Projected Requirements*[1] | | | | | | | | |
| Gas—375 $10^9$ $m^3$ | 1970 | | | 540 | | | 2,420 | | |
| | 1975 | | | 1,010 | | | 4,860 | | |
| | 1980 | | | 2,060 | | | 9,800 | | |
| *Dahomey* | 1959 | — | 3·2 | 3·2 | — | 7·5 | 7·5 | 3·3 | |
| Energy Reserves | 1965 | — | 12 | 12 | — | 21 | 21 | 7·7 | 79 |
| Hydro—3,000 Gwh | 1959–65 Annual Increase | | | | | .. | 20·0% | | |
| Oil— | *Projected Requirements*[1] | | | | | | | | |
| Gas— | 1970 | | | 24 | | | 48 | | |
| | 1975 | | | 34 | | | 90 | | |
| | 1980 | | | 60 | | | 170 | | |
| *Togo* | 1959 | — | 3·1 | 3·1 | — | 3·4 | 3·4 | 1·9 | |
| Energy Reserves | 1965 | 1·6 | 13·4[1] | 15·0 | 3·6 | 31·4[7] | 35·0 | 24·0 | 144 |
| Hydro—2,000 Gwh | 1959–65 Annual Increase | | | | | .. | 37·0% | | |
| Oil— | *Projected Requirements*[1] | | | | | | | | |
| Gas— | 1970 | | | 23 | | | 70 | | |
| | 1975 | | | 40 | | | 140 | | |
| | 1980 | | | 72 | | | 280 | | |
| *Ghana* | 1959 | — | 116 | 116 | — | 365 | 365 | 55·0 | |
| Energy Reserves | 1965 | 512 | 180 | 692[8] | — | 570 | 570 | 72·0 | 462 |
| Hydro—10,500 Gwh | 1959–65 Annual Increase | | | | | .. | 9·2% | | |
| Oil— | *Projected Requirements*[2] | | | | | | | | |
| Gas— | 1970 | | | 562 | | | 3,650 | | |
| | 1975 | | | 990 | | | 6,100 | | |
| | 1980 | | | 1,660 | | | 9,640 | | |
| *Upper Volta* | 1959 | — | 2·7 | 2·7 | — | 7 | 7 | 1·5 | |
| Energy Reserves | 1965 | — | 11·0 | 11·0 | — | 21 | 21 | 3·3 | 53 |
| Hydro—4,000 Gwh | 1959–65 Annual Increase | | | | | .. | 20·3% | | |
| Oil— | *Projected Requirements*[3] | | | | | | | | |
| Gas— | 1970 | | | 18 | | | 35 | | |
| | 1975 | | | 25 | | | 70 | | |
| | 1980 | | | 40 | | | 100 | | |

Table 4—*continued*

| | | | | | | | | | |
|---|---|---|---|---|---|---|---|---|---|
| *Niger* | 1959 | — | 2·4 | 2·4 | — | 6 | 6 | 2·1 | |
| Energy Reserves | 1965 | — | 7·3 | 7·3 | — | 19 | 19 | 4·7 | 56 |
| Hydro— 200 Gwh | 1959–65 Annual Increase | | | | | | .. | 19·1% | |
| Oil— | *Projected Requirements*[1] | | | | | | | | |
| Gas— | 1970 | | | 15 | | | 38 | | |
| | 1975 | | | 25 | | | 77 | | |
| | 1980 | | | 45 | | | 154 | | |
| *Mali* | 1959 | 0·5 | 5·0 | 5·5 | 1 | 13 | 14 | 3·5 | |
| Energy Reserves | 1965 | 0.5 | 15·3 | 16 | 1 | 31 | 32 | 5·3 | 112 |
| Hydro— 20,000 Gwh | 1959–65 Annual Increase | | | | | .. | 9·8% | | |
| Oil— | *Projected Requirements*[1] | | | | | | | | |
| Gas— | 1970 | | | 30 | | | 64 | | |
| | 1975 | | | 45 | | | 129 | | |
| | 1980 | | | 75 | | | 260 | | |
| *Ivory Coast* | 1959 | 20 | 14 | 34 | 5 | 47 | 52 | 14·4 | |
| Energy Reserves | 1965 | 50 | 50 | 100 | 141 | 79 | 220 | 46·5 | 568 |
| Hydro— 20,000 Gwh | 1959–65 Annual Increase | | | | | .. | 27·1% | | |
| Oil— | *Projected Requirements*[4] | | | | | | | | |
| Gas Gas— | 1970 | | | 124 | | | 510 | | |
| | 1975 | | | 275 | | | 1,100 | | |
| | 1980 | | | 480 | | | 1,990 | | |
| *Liberia* | 1959 | — | 20 | 20 | 20 | 60 | 80 | 80 | |
| Energy Reserves | 1965 | 38[9] | 114 | 152[9] | 20 | 255 | 275 | 260 | 90 |
| Hydro— 25,000 Gwh | 1959—65 Annual Increase | | | | | .. | 22·4% | | |
| Oil— | *Projected Requirements*[4] | | | | | | | | |
| Gas— | 1970 | | | 180 | | | 530 | | |
| | 1975 | | | 280 | | | 1,070 | | |
| | 1980 | | | 500 | | | 1,970 | | |
| *Sierra Leone* | 1959 | — | 21 | 21 | — | 40 | 40 | 17 | |
| Energy Reserves | 1965 | — | 53 | 53 | — | 94 | 94 | 41 | 358 |
| Hydro— 10,000 Gwh | 1959–65 Annual Increase | | | | | .. | 16·9% | | |
| Oil— | *Projected Requirements*[5] | | | | | | | | |
| Gas— | 1970 | | | 60 | | | 151 | | |
| | 1975 | | | 100 | | | 244 | | |
| | 1980 | | | 165 | | | 392 | | |
| *Guinea* | 1959 | n.a. | n.a. | | 18 | 11 | 29 | 9·0 | |
| Energy Reserves | 1965 | 21 | 55 | 76 | 43 | 134 | 177 | 47 | 108 |
| Hydro— 25,000 Gwh | 1959–65 Annual Increase | | | | | | 35·6% | | |
| Oil— | *Projected Requirements*[6] | | | | | | | | |
| Gas— | 1970 | | | 150 | | | 356 | | |
| | 1975 | | | 450 | | | 2,375 | | |
| | 1980 | | | 620 | | | 3,100 | | |

Table 4—*continued*

| | | | | | | | | | |
|---|---|---|---|---|---|---|---|---|---|
| *Gambia* | 1959 | — | 2·9 | 2·9 | — | 4·2 | 4·2 | 13 | |
| Energy Reserves | 1965 | — | 3·5 | 3·5 | — | 7·0 | 7·0 | 19 | 20 |
| Hydro— | | | | | | | | | |
| | 1959–65 Annual Increase | | | | | .. | 7·9% | | |
| Oil— | *Projected Requirements*[5] | | | | | | | | |
| Gas— | 1970 | | | 6 | | | 11 | | |
| | 1975 | | | 10 | | | 18 | | |
| | 1980 | | | 20 | | | 29 | | |
| *Senegal* | 1959 | — | 50 | 50 | — | 185 | 185 | 54·5 | |
| Energy Reserves | 1965 | — | 100 | 100 | — | 317 | 317 | 79·0 | 730 |
| Hydro— | | | | | | | | | |
| 16,000 Gwh | 1959–65 Annual Increase | | | | | .. | 9·2% | | |
| Oil— | *Projected Requirements*[1] | | | | | | | | |
| Gas— | 1970 | | | 180 | | | 638 | | |
| | 1975 | | | 314 | | | 1,280 | | |
| | 1980 | | | 540 | | | 2,580 | | |
| *Mauritania* | 1959 | n.a. | n.a. | | — | 1.2 | 1·2 | 0·3 | |
| Energy Reserves | 1965 | — | 21 | 21 | — | 43·1 | 43·1[10] | 48·0 | 32 |
| Hydro— | | | | | | | | | |
| 200 Gwh | 1959–65 Annual Increase | | | | | .. | 85% | | |
| Oil— | *Projected Requirements*[1] | | | | | | | | |
| Gas— | 1970 | | | 40 | | | 87 | | |
| | 1975 | | | 60 | | | 174 | | |
| | 1980 | | | 105 | | | 351 | | |
| *Total West Africa* | 1959 | 41 | 443 | 484 | 91 | 1,068 | 1,159 | | |
| Energy Reserves | 1965 | 651 | 938 | 1,589 | 334 | 2,704 | 3,038 | | 4,272 |
| Hydro— | | | | | | | | | |
| 155,400 Gwh | 1959–65 Annual Increase | | | | | .. | 17·5% | | |
| Oil— | *Projected Requirements:* | | | | | | | | |
| Gas— | 1970 | | | 1,952 | | | 8,608 | | 5,562 |
| | 1975 | | | 3,658 | | | 17,727 | | 7,410 |
| | 1980 | | | 6,442 | | | 30,816 | | 8,948 |

NOTES:

[1] Future requirements estimated by projecting energy production at 15 per cent per annum from 1965.

[2] Future requirements estimated by projecting energy production at 15 per cent per annum from 1965 plus an assumed allowance for the aluminium industry:

300 MW in 1970
450 MW in 1975
600 MW in 1980 (at 95 per cent plant factor).

[3] Future requirements estimated by projecting energy production at 13 per cent per annum from 1965.

[4] Future requirements extracted from Cavally River Project Preliminary Report to A.D.B. dated November 1966. As Liberia is the only 60-cycle system in West Africa, the intertie transmission system with the Ivory Coast must accommodate for frequency conversion.

Table 4—*continued*

[5] Future requirements estimated by projecting energy production at 10 per cent per annum from 1965.

[6] Future requirements estimated by projecting energy production at 15 per cent per annum from 1965 plus an assumed allowance for an aluminium industry: 200 MW from 1975.

[7] Includes 8·6 MW at the phosphate mine which is assumed supplied by less expensive public power by 1970. Load growth may be conservative when domestic chemical industry develops.

[8] Includes 100 MW in public sector and 80 MW in private sector, nearly all of which is diesel. In October 1965 first units of 512 MW installation at Akosombo came on line.

[9] Includes 34 MW Mt. Coffee station commissioned in 1966.

[10] Includes an allowance for production and consumption by MIFERMA iron ore mine.

TABLE 5

CAPITAL REQUIREMENTS FOR WEST AFRICAN COUNTRIES ELECTRIC POWER SYSTEMS, 1966–70, 1971–75 AND 1976–80

| Country and Fvie-Year Period | New Power Generation Requirements: Diesel or Thermal Installed Capacity Mw | Diesel or Thermal Cost $ Million | Hydro-electric Installed Capacity Mw | Hydro-electric Cost $ Million | Total Capacity Mw | Total Cost $ Million | New H.T. Transmission Requirements: Distance, Miles | Capacity, Kv | Cost $ Million | Power Supply System Total Cost $ Million | Description of Hydro-electric Developments |
|---|---|---|---|---|---|---|---|---|---|---|---|
| *Nigeria* | | | | | | | | | | | |
| 1966–70 | 208[1] | 39·6 | 320 | 185·0[2] | 528 | 224·6 | 810 | 330 | 30·0 | 261·6[13] | Kainji first stage |
| | | | | | | | 340 | 132 | 7·0 | | |
| 1971–75 | — | | 640 | 32·0 | 640 | 32·0 | — | — | — | 32·0 | Kainji second stage |
| 1976–80 | — | | 980 | 188·0 | 980 | 188·0 | 60 | 330 | 2·7 | 190·7 | Jebba and Shiroro Gorge |
| 15-year Total | 208 | 39·6 | 1,940 | 405·0 | 2,148 | 444·6 | 1,210 | | 39·7 | 484·3 | |
| *Dahomey* | | | | | | | | | | | |
| 1966–70 | — | — | [3] | — | — | — | — | — | — | — | |
| 1971–75 | — | — | — | — | — | — | — | — | — | — | |
| 1976–80 | — | — | 700 | 175·0 | 700 | 175·0 | 60 | 161 | 1·6 | 176·6 | Oueme Development |
| 15-year Total | | | 700 | 175·0 | 700 | 175·0 | 60 | | 1·6 | 176·6 | |
| *Togo* | | | | | | | | | | | |
| 1966–70 | | | [4] | — | — | — | — | — | — | — | |
| 1971–75 | | | 60 | 64·0 | 60 | 64·0 | 20 | 161 | 0·5 | 64·5 | Mono River First Development |
| 1976–80 | | | 55 | 48·0 | 55 | 48·0 | 50 | 161 | 1·3 | 49·3 | Mono River Second Development |
| 15-year Total | | | 115 | 112·0 | 115 | 112·0 | 70 | | 1·8 | 113·8 | |
| *Ghana* | | | | | | | | | | | |
| 1966–70 | 2 | 0·3 | — | — | 2 | 0·3 | 140[8] | 161 | 2·4 | 2·7 | |
| 1971–75 | | | 342 | 36·4 | 342 | 36·4 | — | | — | 36·4 | Akosombo second stage plus Kpong |
| 1976–80 | | | 119[5] | 35·7 | 119 | 35·7 | 400 | 161 | 10·0 | 45·7 | Bui Development |
| 15-year Total | 2 | 0·3 | 461 | 72·1 | 463 | 72·4 | 540 | | 12·4 | 84·8 | |
| *Upper Volta* | | | | | | | | | | | |
| 1966–70 | 10 | 1·5 | | — | 10 | 1·5 | — | | — | 1·5 | |
| 1971–75 | | | 10[6] | 4·0 | 10 | 4·0 | 320 | 90 | 6·4 | 10·4 | Dedougou & Banfora Developments |
| 1976–80 | | | | | — | — | 120 | 161 | 3·0 | 3·0 | |
| 15-year Total | 10 | 1·5 | 10 | 4·0 | 20 | 5·5 | 440 | | 9·4 | 14·9 | |
| *Niger* | | | | | | | | | | | |
| 1966–70 | 9 | 1·4 | — | — | 9 | 1·4 | — | — | — | 1·4 | |
| 1971–75 | | | 33 | 13·2 | 33 | 13·2 | — | — | — | 13·2 | Niamey Development |
| 1976–80 | — | — | — | — | — | — | — | — | — | | |
| 15-year Total | 9 | 1·4 | 33 | 13·2 | 42 | 14·6 | — | | — | 14·6 | |
| *Mali* | | | | | | | | | | | |
| 1966–70 | 20 | 3·0 | — | — | 20 | 3·0 | — | | — | 3·0 | |
| 1971–75 | | | 300 | 75·0 | 300 | 75·0 | 240 | 225 | 7·2 | 86·7 | Gouina Development |
| | | | | | | | 100 | 330 | 4·5 | | |
| 1976–80 | — | — | — | — | — | — | — | — | — | — | |
| 15-year Total | 20 | 3·0 | 300 | 75·0 | 320 | 78·0 | 340 | | 11·7 | 89·7 | |

| | | | | | | | | | | | |
|---|---|---|---|---|---|---|---|---|---|---|---|
| *Ivory Coast* | | | | | | | | | | | |
| 1966–70 | 32 | 7·7 | — | — | 32 | 7·7 | — | | — | 7·7 | |
| 1971–75 | | | 174 | 52·0 | 174 | 52·0 | — | 225 | 11·8 | 63·8 | Kossou Development |
| 1976–80 | | | 150 | 38·0 | 150 | 38·0 | 286 | | 12·4 | 100·4 | (Tiboto) and Brimbo Developments |
| | | | (360)[7] | (100·0) | (360) | (100·0) | | 330 | | | |
| 15-year Total | 32 | 7·7 | 504 | 140·0 | 536 | 147·7 | 286 | | 24·2 | 171·9 | |
| *Liberia* | | | | | | | | | | | |
| 1966–70 | | | 68 | 18·7 | 68 | 18·7 | — | | — | 18·7 | Mt. Coffee Upstream Storage |
| 1971–75 | 60 | 14·4 | — | — | 60 | 14·4 | — | | — | 14·4 | |
| 1976–80 | | | (360)[7] | (100·0) | (360) | (100·0) | 187 | 330 | 8·4 | 58·4 | (Tiboto Development) |
| 15-year Total | 60 | 14·4 | 248 | 68·7 | 308 | 83·1 | 187 | | 8·4 | 91·5 | |
| *Sierra Leone* | | | | | | | | | | | |
| 1966–70 | 7 | 1·1 | — | — | 7 | 1·1 | — | | — | 1·1 | |
| 1971–75 | | | 42 | 16·8 | 42 | 16·8 | 50 | 225 | 1·5 | 18·3 | Little Scarcies (Mange) |
| 1976–80 | | | 63 | 26·4 | 63 | 26·4 | 75 | 225 | 2·3 | 28·7 | Seli (Bumbuna) |
| 15-year Total | 7 | 1·1 | 105 | 43·2 | 112 | 44·3 | 125 | | 3·8 | 48·1 | |
| *Guinea* | | | | | | | | | | | |
| 1966–70 | 80 | 12·0 | — | — | 80 | 12·0 | — | | — | 12·0 | |
| 1971–75 | | | 480 | 120·0 | 480 | 120·0 | 170 | 225 | 5·1 | 125·1 | Souapiti |
| 1976–80 | | | 240 | 60·0 | 240 | 60·0 | 300 | 225 | 9·0 | 69·0 | Amaria first stage |
| 15-year Total | 80 | 12·0 | 720 | 180·0 | 800 | 192·0 | 470 | | 14·1 | 206·1 | |
| *Gambia* | | | | | | | | | | | |
| 1966–70 | 5 | 0·8 | —[9] | — | 5 | 0·8 | — | | — | 0·8 | |
| 1971–75 | | | | — | — | — | 25 | 225 | 0·8 | 0·8 | |
| 1976–80 | — | — | — | — | — | — | — | — | — | — | |
| 15-year Total | 5 | 0·8 | | | 5 | 0·8 | 25 | | 0·8 | 1·6 | |
| *Senegal* | | | | | | | | | | | |
| 1966–70 | 100 | 19·0 | — | — | 100 | 19·0 | — | | — | 19·0 | |
| 1971–75 | | | [10] | — | — | — | 362 | 330 | 16·3 | 23·3 | |
| | | | | | | | 232 | 225 | 7·0 | | |
| 1976–80 | | | 100 | 30·0 | 100 | 30·0 | | | — | 45·0 | Gambia River Development 100 MW |
| | | | (100)[11] | (30·0) | (100) | (30·0) | | | | | (Senegal River at Bakel shared with Mauritania) |
| 15-year Total | 100 | 19·0 | 150 | 45·0 | 250 | 64·0 | 594 | | 23·3 | 87·3 | |
| *Mauritania* | | | | | | | | | | | |
| 1966–70 | 20 | 3·0 | — | — | 20 | 3·0 | — | | — | 3·0 | |
| 1971–75 | | | — | — | — | — | 328 | 225 | 9·8 | 9·8 | |
| 1976–80 | | | (100)[12] | (30·0) | (100) | (30·0) | — | | — | 15·0 | Bakel Development shared with Senegal |
| 15-year Total | 20 | 3.0 | 50 | 15·0 | 70 | 18·0 | 328 | | 9·8 | 27·8 | |
| *Total West Africa* | | | | | | | | | | | |
| 1966–70 | 493 | 89·4 | 388 | 203·7 | 861 | 293·1 | 1,290 | | 39·4 | 332·5 | |
| 1971–75 | 60 | 14·4 | 2,081 | 413·4 | 2,141 | 427·8 | 1,847 | | 70·9 | 498·7 | |
| 1976–80 | — | — | 2,867 | 731·1 | 2,867 | 731·1 | 1,538 | | 50·7 | 781·8 | |
| 15-year Total | 553 | 103·8 | 5,336[14] | 1,348·2[14] | 5,889 | 1452·0 | 4,675[14] | | 161·0[14] | 1,6131·0[14] | |

Calculation Note:

The order of magnitude capital costs shown in this table were obtained where possible from engineers estimated costs. Where these were not available capital costs were estimated using the following assumptions:

1. Thermal Generating Stations:
   a. Diesel—up to 20 MW estimated at $150 per KW
   b. Steam—30 MW approximate size estimated at $240 per KW
   50 MW approximate size estimated at $190 per KW
2. Hydro-electric Generating Stations:
   Stations below 75 MW installed capacity estimated at $400 per KW
   ,, 75 MW–200 MW ,, ,, ,, ,, $300 per KW
   ,, over 200 MW ,, ,, ,, ,, $250 per KW
3. Transmission:
   330 Kv power lines estimated at $45,000 per mile
   225 Kv ,, ,, ,, ,, $30,000 per mile
   90 Kv ,, ,, ,, ,, $20,000 per mile

These costs relate to construction, equipment and installation costs and do not include engineering services or interest during construction, which usually each total around 10 per cent of costs.

Notes:

[1] This thermal power installation includes a variety of small and large stations: including 178 MW of gas turbine and ·30 MW of diesel.

[2] Does not include an estimated $22 million for navigation locks.

[3] Dahomey will purchase excess power from Akosombo under the scheme now being examined for Togo and Dahomey. As the Akosombo and additional power facilities in Ghana are fully utilised Dahomey will purchase power from the Togo Mono Development and finally supply her own power from the Oueme Development.

[4] Togo along with Dahomey to purchase power from Ghana until the Mono Development is commissioned.

[5] By 1980 Ghana will be requiring further power which will become available from the Oueme Development in Dahomey.

[6] By 1980 Upper Volta will be interconnected with Ghana, Dahomey and Togo and its small estimated demand may be satisfied from this grid.

[7] This figure of 360 MW shows the whole of the installed capacity at Tiboto which is assumed shared evenly by Ivory Coast and Liberia. However, to ease the shortage of power in Liberia an arrangement might be agreed with the Ivory Coast to use additional power from Tiboto.

[8] Includes 75 miles 161 Kv international link from Prestea to Ayame required by 1970 to allow Ivory Coast to meet 1971 and 1972 requirements until the construction of Kossou, and future load interchanges.

[9] Gambia, having a small total requirement even in 1980 should purchase some of the power available from Mali.

[10] Senegal having limited power facilities so far envisaged will require to purchase power from Mali from 1975 onwards.

[11] This figure includes 100 MW which will be installed in Senegal and 100 MW which will be installed on the Senegal River between Senegal and Mauritania. Senegal will be entitled to 50 MW of this installation. After 1975 Mauritania will probably require to purchase from Mali.

[12] This figure refers to the Bakel Development, half of which will be available to Mauritania. Mauritania will import power from Mali from 1975.

[13] Transmission costs and generating station costs have been included for the first stage of the new Kainji system although this system is now under construction and finance, therefore, secured.

[14] Excluding the Kainji project under construction the total new hydro-electric requirements are 5,016 MW and $1,163 million and transmission lines of 3,526 miles costing $124 million.